Process Development Division 2023

AA009203

Held at the 2023 AIChE Spring Meeting and 19th Global Congress on Process Safety

Houston, Texas, USA
12-16 March 2023

ISBN: 978-1-7138-7928-2

Printed from e-media with permission by:

Curran Associates, Inc.
57 Morehouse Lane
Red Hook, NY 12571

proceedings.com

Some format issues inherent in the e-media version may also appear in this print version.

Printed with permission by Curran Associates, Inc. (2023)

For permission requests, please contact AIChE
at the address below.

AIChE
120 Wall Street, FL 23
New York, NY 10005-4020

Phone: (800) 242-4363
Fax: (203) 775-5177

www.aiche.org

Additional copies of this publication are available from:

Curran Associates, Inc.
57 Morehouse Lane
Red Hook, NY 12571 USA
Phone: 845-758-0400
Fax: 845-758-2633
Email: curran@proceedings.com
Web: www.proceedings.com

TABLE OF CONTENTS

PROCESS SCALE UP AND SAFETY

MODELING TOOLS AND TECHNIQUES FOR PROCESS RANDD II

ADVANCES IN PROCESS DESIGN AND DEVELOPMENT II

118a Decarbonization Roadmap Methodology and Case Study for Petrochemical Facilities

David Houser, Omar Hamid

The petrochemical industry is a large emitter of carbon dioxide but is a "difficult-to-decarbonize" sector. The life cycle emissions are attributed to a combination of sources including upstream, fuel-related, electricity generation, industrial process and downstream product use. Achieving net-zero is an ambitious goal that requires careful planning and implementation.

The paper explores corporate decarbonate strategies to achieve 2030 and 2050 goals. Energy efficiency, electrifications, feedstock switching and CCUS are investigated. A methodology is outlined to develop unique approaches for companies and facilities. Practical technologies will be reviewed and a process to deem their applicability to specific sites is discussed. A sample marginal cost abate curves is shared. A methodology to develop and optimized implementation schedule is discussed.

118b Hydrodeoxygenation of Bio-Oil Via in-Situ Alkene Dehydrogenation

Abdullah Baubaid, Marcos Millan

The abundance of waste biomass and its extracted lignin and its intrinsic aromatic network could provide the chemical industry with a rich, sustainable, and renewable hydrocarbon supply. This potential source of green added-value chemicals, like benzene, toluene and xylenes (BTX) boosts the interest on biorefinery and its integration with the existing petrochemical state-of-the-art processes. With the relatively high oxygen content in bio-oil, which hinders its heating value and processability, hydrodeoxygenation (HDO) becomes a crucial step in biorefineries [1], [2]. Conventionally, HDO is processed using high-pressure (up to 200 bar) expensive hydrogen [1]. Consequently, the use of a hydrogen donor is investigated as an alternate strategy by using solvent and/or co-feeding at different operating conditions [3]–[5]. Optimally, the reaction used to produce hydrogen should in itself also lead to added-value byproducts. Hence, dehydrogenation of alkanes can be an interesting choice as valuable alkenes would be generated.

This research aims to demonstrate HDO of bio-oil via in-situ alkane dehydrogenation. Alkanes such as ethane and propane are the primary feeds of the petrochemical value chain to produce the building block chemicals ethylene, propylene, and aromatics. Therefore, the incorporation of alkenes production with upgraded bio-oil would improve the overall outlook of HDO. This research seeks to increase bio-oil conversion due to the catalytic formation of reactive hydrogen at the catalyst surface derived from the dehydrogenation of alkenes. In addition, this research seeks to increase alkane conversion derived from the continuous consumption of hydrogen for bio-oil upgrading. To investigate alkane and bio-oil conversions, experiments are conducted on a single catalytic system comprised of HDO and alkane dehydrogenation catalysts in a continuous up-flow reactor. To achieve higher hydrogen formation from thermodynamic equilibrium perspective, these experiments are conducted at atmospheric pressure and high temperature environment using guaiacol as model compound. Catalytically, these studies were conducted on Ni-Mo and MoO_3 based catalyst for HDO, and Cr-based catalyst for alkane dehydrogenation [1], [6]–[9].

A bio-oil conversion similar to that achieved through the use of hydrogen was observed, with products having a higher C/O ratio, as well as ethylene production, when using alkane dehydrogenation catalysts and a superior conversion with HDO catalysts. In addition, increases in reaction temperature and catalyst loading have a significant positive effect on HDO performance. On the other hand, catalytic dehydrogenation of ethane suffers a decline in conversion after 30 minutes of time-on-stream until it reaches a plateau; the introduction of bio-oil model compounds accelerates this decline through catalyst deactivation, necessitating frequent of catalyst re-activation. In summary, the results demonstrated the feasibility of replacing hydrogen with alkanes and there is scope by further developing the process through more effective catalyst formulations.

References

[1] E. Furimsky, "Catalytic hydrodeoxygenation," *Appl. Catal. A Gen.*, vol. 199, no. 2, pp. 147–190, Jun. 2000, doi: 10.1016/S0926-860X(99)00555-4.

[2] Z. Si, X. Zhang, C. Wang, L. Ma, and R. Dong, "An Overview on Catalytic Hydrodeoxygenation of Pyrolysis Oil and Its Model Compounds," *Catal. 2017, Vol. 7, Page 169*, vol. 7, no. 6, p. 169, Jun. 2017, doi: 10.3390/CATAL7060169.

[3] A. T. To and D. E. Resasco, "Hydride transfer between a phenolic surface pool and reactant paraffins in the catalytic cracking of m-cresol/hexanes mixtures over an HY zeolite," *J. Catal.*, vol. 329, pp. 57–68, Sep. 2015, doi: 10.1016/J.JCAT.2015.04.025.

[4] M. E. Domine, A. C. Van Veen, Y. Schuurman, and C. Mirodatos, "Coprocessing of Oxygenated Biomass Compounds and Hydrocarbons for the Production of Sustainable Fuel," *ChemSusChem*, vol. 1, no. 3, pp. 179–181, Mar. 2008, doi: 10.1002/CSSC.200700049.

[5] X. dong Fan *et al.*, "Benzene, toluene and xylene (BTX) from in-situ gas phase hydrodeoxygenation of guaiacol with liquid hydrogen donor over bifunctional non-noble-metal zeolite catalysts," *Renew. Energy*, vol. 152, pp. 1391–1402, Jun. 2020, doi: 10.1016/j.renene.2020.01.015.

[6] U. Olsbye, A. Virnovskaia, O. Prytz, S. J. Tinnemans, and B. M. Weckhuysen, "Mechanistic Insight in the Ethane Dehydrogenation Reaction over Cr/Al2O3 Catalysts," *Catal. Lett. 2005 1031*, vol. 103, no. 1, pp. 143–148, Sep. 2005, doi: 10.1007/S10562-005-6521-7.

[7] B. M. Weckhuysen and R. A. Schoonheydt, "Alkane dehydrogenation over supported chromium oxide catalysts," *Catal. Today*, vol. 51, no. 2, pp. 223–232, Jun. 1999, doi: 10.1016/S0920-5861(99)00047-4.

[8] G. Zhou, P. A. Jensen, D. M. Le, N. O. Knudsen, and A. D. Jensen, "Atmospheric Hydrodeoxygenation of Biomass Fast Pyrolysis Vapor by MoO3," *ACS Sustain. Chem. Eng.*, vol. 4, no. 10, pp. 5432–5440, Oct. 2016, doi: 10.1021/acssuschemeng.6b00757.

[9] I. D. Mora-Vergara, L. Hernández Moscoso, E. M. Gaigneaux, S. A. Giraldo, and V. G. Baldovino-Medrano, "Hydrodeoxygenation of guaiacol using NiMo and CoMo catalysts supported on alumina modified with potassium," *Catal. Today*, vol. 302, pp. 125–135, Mar. 2018, doi: 10.1016/j.cattod.2017.07.015.

118c Critical Role of Surface Hydroxyl-Amine Interactions in Direct Air Capture Using Single Component and Composite PEI-Silica Sorbents

Xiaowei Wu, Ramanan Krishnamoorti, Praveen Bollini

Polyethyleneimine (PEI)-impregnated silica adsorbents have been widely explored as candidates for direct air capture, and the co-impregnation of polyethylene glycol (PEG) is a commonly used approach for improving amine efficiency. Improvements in amine utilization from PEG addition were previously rationalized mainly from changes in CO_2 binding chemistry, amine-amine cooperative interactions, and the mitigation of diffusion limitations. We present herein evidence for the determinative role of amine-surface hydroxyl interactions in controlling CO_2 sorption behavior in the PEI-silica system. PEG addition appears to disrupt surface hydroxyl-amine interactions to release "inaccessible" surface-bounded amine sites to free PEI. Unfavorable surface hydroxyl-amine interactions can also be mitigated through surface silylation with hexamethyldisilazane (HMDS) which is, however, not effective when PEG is present, suggesting the analogous role of PEG co-impregnation and HMDS-capping in PEI population transfer for improving amine utilization. Thermal degradation behaviors and FTIR spectra also supports the change of PEI morphology after PEG introduced. Insights gained into the role of amine-surface hydroxyl interactions provide guidelines for the advanced design of aminosilica sorbents for direct air capture applications.

Development and Modeling of Energy-Optimum Aromatic Separation Processes Using Ionic Liquid Solvents

Adam McNeeley[1], Chang-Che Tsai[2], Shiang-Tai Lin[2] and Y. A. Liu[1,]*

**Corresponding author: design@vt.edu; (540) 231-7800.*

[1] AspenTech Center of Excellence in Process System Engineering, Department of Chemical Engineering, Virginia Polytechnic Institute and State University, Blacksburg, Virginia 24061

[2] Department of Chemical Engineering, National Taiwan University, Taipei, Taiwan 10617

1. Introduction

Ionic liquid (IL) solvents are attractive solvents for selective separation of aromatics and aliphatics in hydrocarbon mixtures. There is incentive to extract aromatics from ethylene cracker naphtha feeds, where the current aromatic extraction technologies are not employed. ILs are considered attractive because of their non-volatility allowing easy separation from extract along with the vast number of synthesizable ILs that offer different chemical characteristics.

One of our goals is to quantify the potential of IL solvent in aromatic extraction by understanding the best way to simulate and model an IL to capture interactions between an IL and hydrocarbons. We test this by evaluating how process simulation using *a priori* thermodynamic method COSMO-SAC compares to an experimental data supported UNIQUAC model. We also evaluate how feed simplifications can change design conclusions. Another goal is to develop an energy-optimum process that can meet high purity requirements for aromatic and aliphatic products and take full advantage of the non-volatility of the IL solvent. Our final goal is to use our novel aromatic extraction process design to evaluate 16 well-studied ILs to understand how they compare to an equivalent sulfolane process and identify key IL solvent properties that have the greatest impact on process energy.

2. Effect of Feed Simplification and COSMO-SAC on Process Results

The aromatic extraction process inherently relies on non-ideal molecular interactions to facilitate the selective removal of aromatics from hydrocarbon mixtures, which is impossible to achieve through conventional distillation methods due to the close boiling nature of aromatics to certain aliphatic species, and the existence of azeotropes. Thus, activity coefficient thermodynamic models must be employed to accurately capture the changes in vapor-liquid equilibrium and the occurrence of liquid-liquid equilibrium.

Traditionally, activity coefficient models rely on experimental data to regress binary coefficients that capture molecular interactions of binary mixtures. We find that the NIST database, ILThermo[1-2], has a large compilation of experimental data with relevant hydrocarbons for aromatic extraction and we can use these data to regress UNIQUAC binary coefficients for 16 ILs listed in Figure 1 with 16 hydrocarbons listed in Table 1. Binary coefficient regression data appear in our associated work[3].

Prior works have made simplifications to the feed composition used in process simulations to reduce the number of hydrocarbon components to consider in the model. We want to

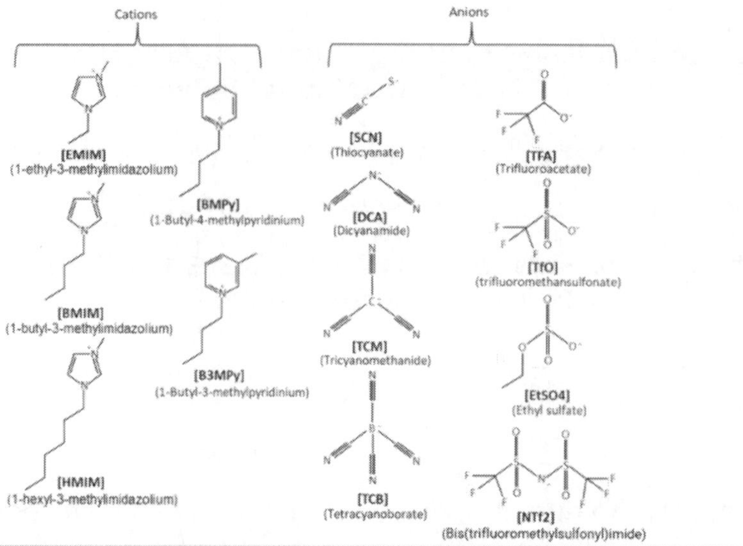

[BMIM][SCN] IL 1	[HMIM][SCN] IL 2	[EMIM][TFA] IL 3	[EMIM][SCN] IL 4	[EMIM][TCM] IL 5	[EMIM][DCA] IL 6	[BMIM][TfO] IL 7	[EMIM][TCB] IL 8
[BMPY][SCN] IL 9	[B3MPY][DCA] IL 10	[BMPY][TCM] IL 11	[B3MPY][TfO] IL 12	[EMIM][NTf2] IL 13	[EMIM][ETSO4] IL 14	[BMPY][NTf2] IL 15	[BMIM][NTf2] IL 16

Figure 1. List of ionic liquid anions and cations considered for the 16 ILs selected in this study. The list of 16 ILs is provided at the bottom of the figure. Adapted from McNeeley et al.[3]

understand how these simplifications impacted the process results and choose two common types of feed simplifications to used in a process model. Table 1 shows the feed simplifications with Case 1 assuming a binary feed of heptane and toluene and case 2 being a feed excluding naphthenes and residual light and heavy components. Case 3 is the original ethylene cracker naphtha feed composition provided by Meindersma[4]. The details of the process model used for this analysis appear in our associated work[5] and is similar to the configuration shown in Figure 6. We us the IL [EMIM][NTf2] for this comparative study.

Table 1. Feed compositions considered for feed simplifications of case 1 and case 2. Case three represents the full feed composition of a 10wt% aromatic content ethylene cracker naphtha feed presented by Meindersma[4]. Adapted from Tsai et al.[5]

Component	Designation	Case 1 (wt%)	Case 2 (wt%)	Case 3 (wt%)
cyclopentane	N5	-		1.5
cyclohexane	N6	-		6.88
methylcyclohexane	N7	-		4.79
1,3-dimethylcyclohexane	N8	-		8.97
n-butane	P4	-		1.5
n-pentane	P5	-		14.46
n-hexane	P6	-	43.68	19.34
n-heptane	P7	90.34	15.76	10.97
n-octane	P8	-	30.91	7.78
n-nonane	P9	-		7.18
n-decane	P10	-		6.98
benzene	B	-	1.79	1.79
toluene	T	9.66	2.99	2.99
ethylbenzene	EB	-	1.99	1.99
p-xylene	PX	-	1.89	1.89
o-xylene	OX	-	1	1

Figure 2 compares some of the key process results for the 3 cases. Overall, we find the binary feed assumption to poorly represent a full-component process. Including the heaviest aromatics in the feed is necessary to get accurate solvent rates. The case 2 feed gives process results fairly close to those of case 3 with full feed composition in terms of solvent rate and process energy demand, however neglecting light components leads to drastic underprediction of the operating pressure of the Extractive Distillation Column (EDC). This causes operational differences in the control strategy of the EDC. Neglecting heavy components may lead to the false impression that the separation is feasible, when a neglected heavy component may be unable to be distilled light relative to benzene in the EDC.

Another challenge with modeling IL processes noted in literature is the limited available data for many ILs and the large number of hydrocarbon species present in an aromatic extraction process. For this reason, *a priori* methods, particularly COSMO (COnductor-like

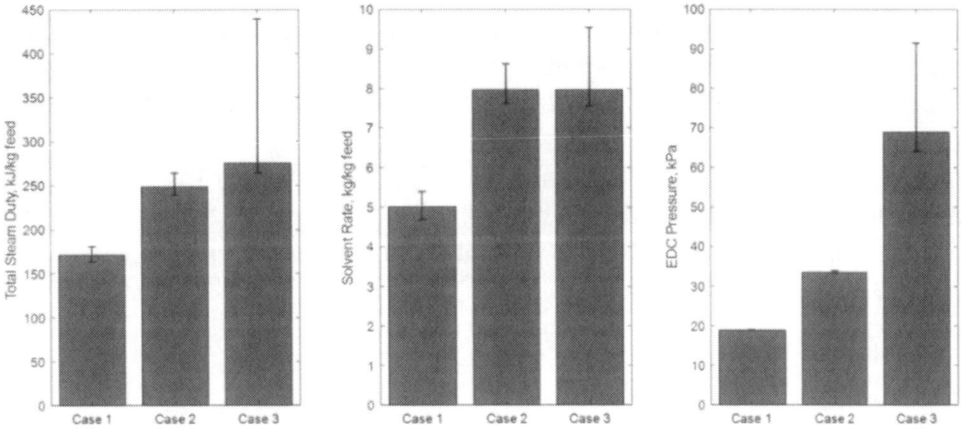

Figure 2. Comparison of major process variables process steam duty, solvent rate, and extractive distillation column (EDC) pressure for the aromatic extraction process using different feed simplifications of an optimized IL aromatic extraction process using [EMIM][NTf2]. Adapted from Tsai et al.[5]

Screening MOdels) models have been employed to predict activity coefficients for hydrocarbons in IL mixtures and are popular in some process modeling studies.

We test the suitability of COSMO-SAC (Segment Activity Coefficient) for process modeling using the improved COSMO-SAC model by Hsieh et al.[6] to generate activity coefficient estimates between hydrocarbons and ILs. Additional details of this methodology appear in our associated work[5].

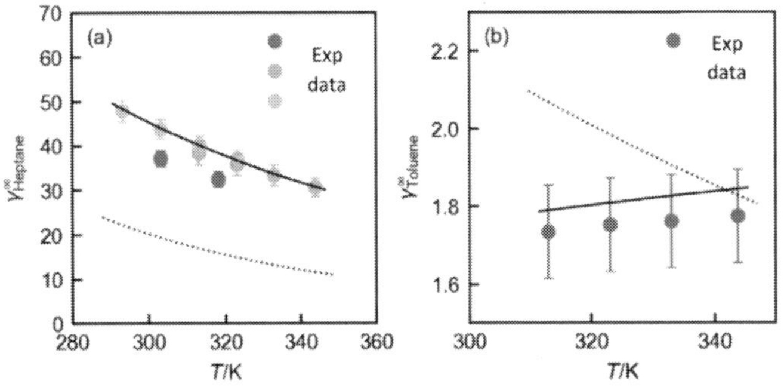

Figure 3. Comparison of UNIQUAC regression (solid line) and COSMO-SAC (dotted line) infinite dilution activity coefficient (IDAC) estimation to experimental IDAC data. Adapted from Tsai et al.[5]

We utilize the IL [EMIM][NTf2] to compare COSMO-SAC predictions to experimental data. The results show that in general, COSMO-SAC can reasonably predict activity coefficients for aromatic components, but consistently underpredicts the activity coefficients of aliphatics. Figure 3 compares the experimental infinite dilution activity coefficient data for heptane with COSMO-SAC predictions and UNIQUAC regression. This trend of underpredicting aliphatic activity coefficient is consistent across vapor-liquid equilibrium and liquid-liquid equilibrium data for all aliphatic species with the order of magnitude of about ½ of experimental values and is present in detail in our associated work[5].

We then quantify the impact of COSMO-SAC on process simulation results using a full aromatic extraction process model. The details of this model are provided in our associated work[5] and resembles the process shown in Figure 6. The impact of the underprediction of activity coefficient by COSMO-SAC impacts the process model primarily in the extractive distillation section where aromatics must be recovered as a heavy product in the bottom of the column and aliphatics as light components in the distillate.

The key separation of this column is then between the lightest aromatic benzene and the heaviest aliphatics. Relative volatility is an effective measure to demonstrate how volatile the key components are relative to each other. We show relative volatilities for heavy aliphatic relative to benzene as infinitely dilute species in [EMIM][NTf2]. The smaller the relative volatility value, the harder the separation and values below 1 mean the aliphatic is heavier than benzene and separation is infeasible to decrease composition of the aliphatic in the extractive distillation column. This ends up being the case for COSMO-SAC for the decane (P10) and nonane (P9) components where the low relative volatility prevents the aromatic purity spec from being met. This contrasts with the binary coefficient regression predictions, which shows the relative volatilities to be approximately double COSMO-SAC predicted values.

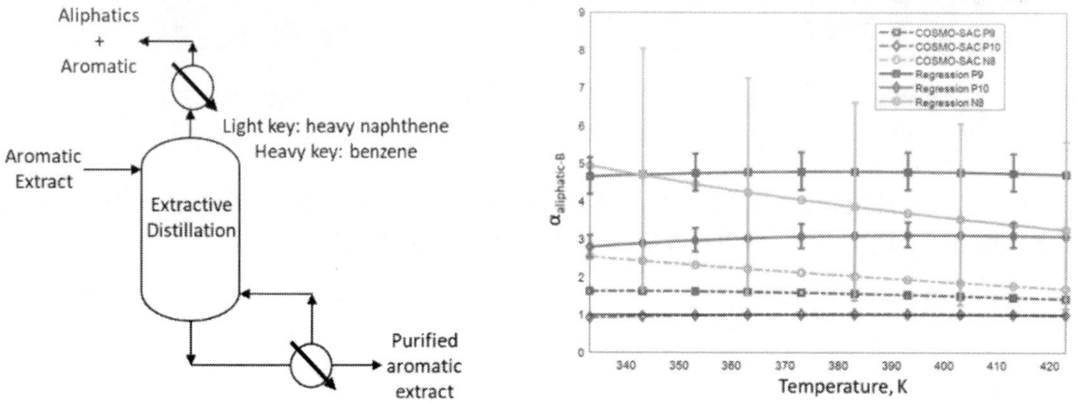

Figure 4. Simple diagram (left) of the extractive distillation and the key separation that determines feasibility of the separation and graph of relative volatility (right) for heavy aliphatic species relative to benzene assuming infinite dilution activity coefficients predicted by UNIQUAC regression (solid line) and COSMO-SAC (dotted line). Adapted from Tsai et al.[5]

3. Development of and Energy-Optimum Process

Another goal of this work is to develop a specific process design tailored to an IL solvent and low aromatic feed compositions of 10-30wt% aromatics. We start by considering the basis of an aromatic liquid extraction process shown in Figure 5. Here, a hydrocarbon mixture is fed into the bottom of a liquid extraction column to separate aromatics from aliphatics and an aromatic free product is produced at the top of the column. The aromatic extract from the bottom of the column goes to an extractive distillation column where aliphatics are stripped from the extract. Finally, the aliphatic free extract mixture goes to a solvent recovery column where aromatics are separated from solvent.

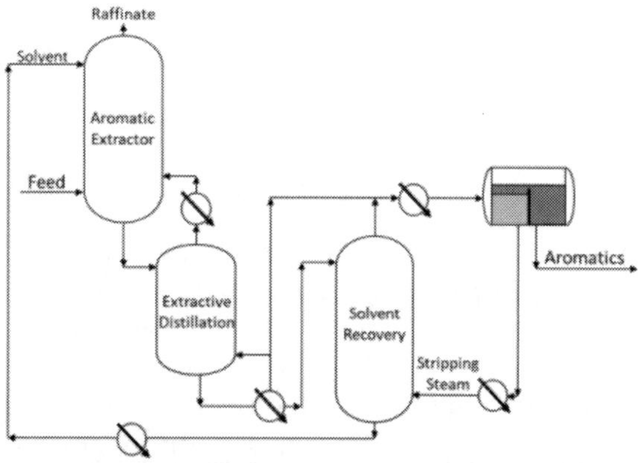

Figure 5. Simple diagram of the basic unit operations required by a aromatic liquid extraction process. Adapted from McNeeley et al.[7]

Using this base process as a starting point, we develop a novel process shown in Figure 6 tailored to the non-volatile property of an IL solvent and low aromatic feed compositions. Details of the synthesis and development of this process are available in our associated work[7]. Overall, we find the claims of ILs allow a fundamentally simpler process to not be realistic, as the only simplification attributed specifically to the non-volatility of ILs compared to the conventional sulfolane solvent is the elimination of the small rectifying section of the SRC and reflux.

A key improvement for a low aromatic feed process design is the independent control of the EDC reboiler temperature. We do this by independently control the vapor rate to the EDC using either a vapor bypass to the SRC or adding stripping steam to the EDC. Another innovation is the improved heat integration network that adds a cross exchanger at the bottom of the EDC to reduce the duty and temperature of the EDC preheater. These improvements enable even sulfolane solvent to be used for low aromatic content feed cases.

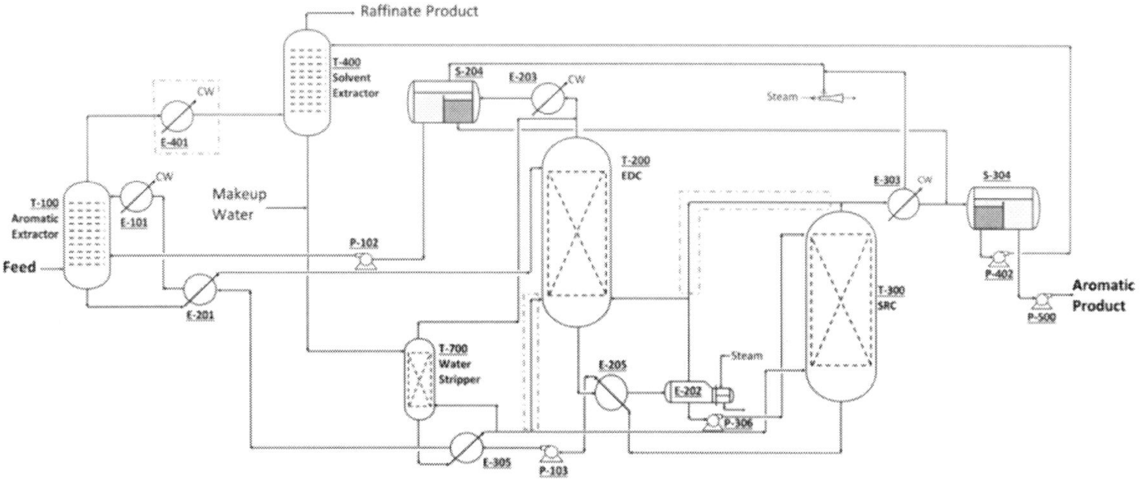

Figure 6. Full process diagram of the novel aromatic extraction process developed in this work. Areas circled in red are equipment or lines that may vary depending on the IL being used. This process can be adapted for sulfolane by adding a rectifying section to the SRC. Adapted from McNeeley et al.[7]

4. Correlation of IL Solvent Variables to Process Performance

We apply our modeling techniques for IL solvents and the novel flowsheet for low aromatic feed contents to 16 ILs and sulfolane. The allows us to create sets of process data that we could then correlate to solvent variables such as selectivity and distribution coefficient. Details about how to define solvent variables are available in our associated work[3].

We introduce a new solvent variable considering our observation that the heat load of the solvent seems to be a dominating factor of the process steam duty and other factors in the process. The heat load of the solvent is effectively the change in enthalpy of the solvent from the minimum temperature to the maximum temperature multiplied by the solvent flow rate as shown in equation 1, where Q_{solv} is the solvent heat load, $\dot{n}_{solv}$ is the solvent flow rate, and Cp_{solv} is the solvent heat capacity. Solvent flow rate is known to be inversely proportional to distribution coefficient, D_i as shown in equation 2. We can also assume that the change in enthalpy of the solvent can be approximated as an average heat capacity at some average temperature when comparing cases that have the same solvent

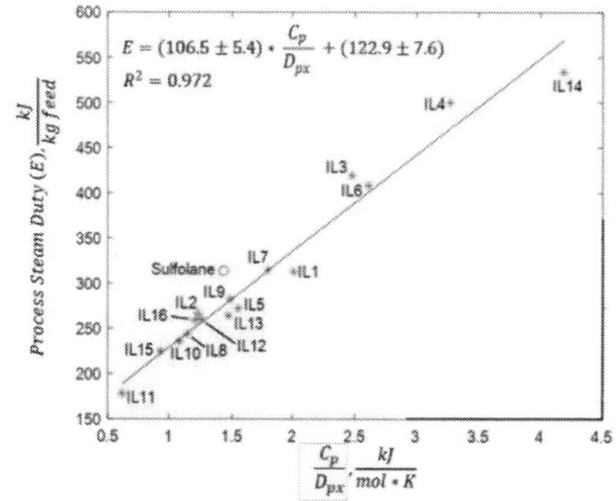

Figure 7. Relationship of process steam duty to solvent heat load variable. Adapted from McNeeley et al.[3]

temperature change as shown in equation 3. This leads us to our new solvent heat load variable in equation 4, which is roughly proportional to the heat load of the solvent.

$$Q_{solv} = \dot{n}_{solv} \int_{T_{min}}^{T_{max}} Cp_{solv}\, dT \qquad (1)$$

$$\dot{n}_{solv} \propto \frac{1}{D_i} \qquad (2)$$

$$\int_{T_{min}}^{T_{max}} Cp_{solv}\, dT \sim \propto Cp_{solv,Tavg} \qquad (3)$$

$$Q_{solv} \sim \propto \frac{Cp_{solv}}{D_i} \qquad (4)$$

We consider 20 different variants of solvent variables that are analyzed for correlation to important process variables such as process steam duty, solvent rate, exchanger 'UA', and other process variables using partial least squares (PLS) analysis. Details of the solvent variables considered and the PLS analysis appear in our associated work[3]. The process variable responsible for almost the entire process operating cost, process steam duty, appears to be highly linearly correlated with our new solvent heat load variable as seen in Figure 7 and is a better correlating parameter than either mass or molar-based distribution coefficient. Additionally, we find that the selectivity to be relatively less important as long as the key separation relative volatility remains above some critical threshold.

We show in Figure 8 that in addition to correlation with process steam duty, the solvent heat load variable also strongly correlates with solvent rate, heat exchanger UA, pump power, and column diameter. This indicates that the total capital cost of an IL aromatic extraction

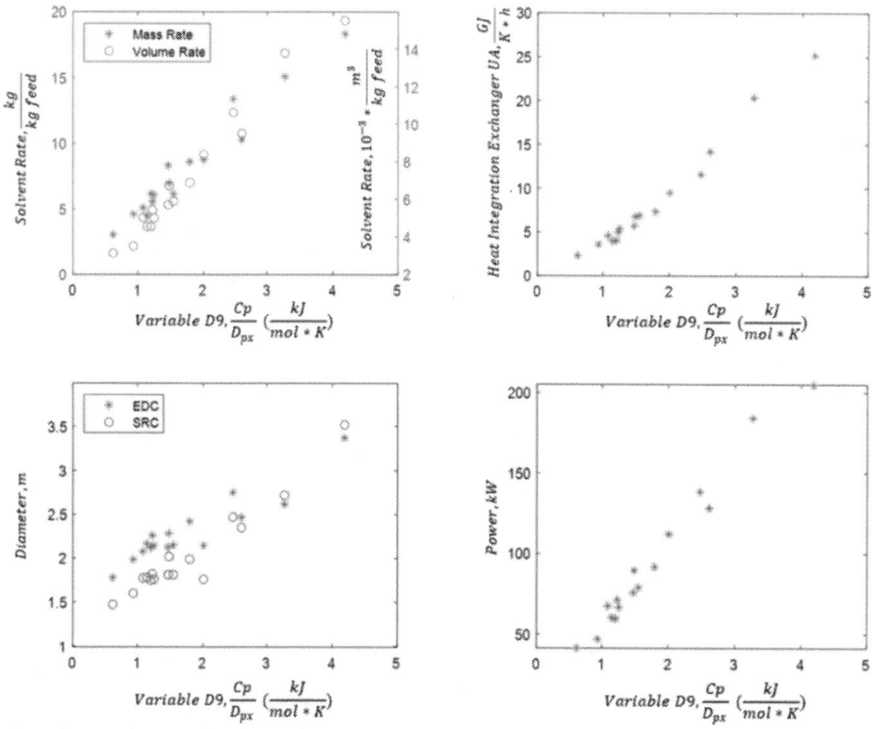

Figure 8. Relationship of capital cost influencing process variables to solvent heat load variable. Adapted from McNeeley et al.[3]

process is likely correlated to the solvent heat load variable, making it a useful objective function to minimize when screening new IL solvents.

5. Conclusion

The primary objective of this work is to develop a comprehensive methodology for evaluating the applicability of ILs as aromatic extraction solvents. We can use publicly available data to capture thermodynamic and molecular interaction properties required to model complex feeds of the process. Using these data, we are then able to develop a novel process tailored to the unique characteristics of an IL solvent as well as the characteristics of a low aromatic content feed. We demonstrate the pitfalls of using feed simplifications or COSMO-SAC activity coefficient prediction as well as the remarkable similarity of an IL aromatic extraction process to conventional aromatic extraction processes.

Upon further analysis of our novel process design, we identify optimizable process variables and operating conditions that minimize total process energy. Some key improvements are the independent control of the maximum solvent operating temperature and improved heat integration.

We then expand the scope of our work to consider 16 well-studied ILs with available experimental data to consider the relationship of IL properties to process performance. We assume that solvent heat load dominates the total energy demand of the process, and thus we introduce a new solvent variable that combines factors both distribution coefficient and heat capacity to give a value that is roughly proportional to the solvent heat load. The results of this study show process steam duty and most of the major capital cost impacting process variables are linearly correlated with this new parameter, which is a better correlating measure than distribution coefficient by itself. It makes it a promising variable to use as an objective function in future screening studies.

Overall, we find ILs as promising alternatives to conventional sulfolane solvent. 11 of the 16 ILs we have tested achieve a lower process steam duty compared to sulfolane with the lowest energy process requiring 57% of the sulfolane steam duty. The relationship between process steam duty and important capital cost influencing process variables mean many of these ILs also offer capital cost advantage. There is still a lot of work to be done before IL solvents can potentially be employed in commercial aromatic extraction applications, but this early evaluation shows ILs have strong potential.

6. References

1. Dong Q., Muzny C.D., Kazakov A., Diky V., Magee J.W., Widegren J.A., Chirico R.D., Marsh K.N., Frenkel M. Ilthermo: A free-access web database for thermodynamic properties of ionic liquids. *Journal of Chemical & Engineering Data*. 2007;52(4):1151-1159.
2. Kazakov A.M., Magee J.W., Chirico R.D., Paulechka E., Diky V., Muzny C.D., Kroenlein K., Frenkel M. Nist standard reference database 147: Nist ionic liquids database - (ilthermo), version 2.0. Accessed 4/1/2022. https://www.nist.gov/srd/nist-standard-reference-database-147
3. McNeeley A., Tsai C.-C., Lin S.-T., Liu Y.A. Science-guided data analytics for selecting ionic solvents for aromatic extraction. *AIChE Journal*. 2023;
4. Meindersma G.W. *Extraction of aromatics from naphtha with ionic liquids*. University of Twente; 2005.
5. Tsai C.-C., McNeeley A., Lin S.-T., Liu Y.A. Evaluation of thermophysical data, cosmo-sac predictions, and feed simplifications for aromatic extraction process simulation using ionic liquid [emim][ntf2]. *AIChE Journal*. 2022:e17916.

6. Hsieh C.-M., Sandler S.I., Lin S.-T. Improvements of cosmo-sac for vapor–liquid and liquid–liquid equilibrium predictions. *Fluid Phase Equilibria*. 2010;297(1):90-97.
7. McNeeley A., Tsai C.-C., Lin S.-T., Liu Y.A. Development of energy-optimum aromatic extraction processes using ionic liquid [emim][ntf2]. *AIChE Journal*. 2022:e17888.

89b Calorimetry Study of Solution Combustion Synthesis of Ni-Nano Particles in a Closed Environment

Anchu Ashok, Luc Vechot, Mamoun Al-Rawashdeh

Solution combustion synthesis (SCS) is a simple, low-cost, and energy-efficient technique for the synthesis of nanomaterials with desired properties suitable for a wide range of applications. Scaling up this synthesis method to an industrial production scale is a very challenging task. The reaction is highly exothermic and releases a significant amount of gas which contains toxic gases. So far, most research works carried out on this route are carried out in an open environment. Containing this chemistry in a closed vessel can help bring it to an industrial production environment. However, it increases the associated hazard due to an accelerated pressure buildup in the closed environment. Calorimetry is a well-established laboratory tool to measure the thermal events associated with material phase change or chemical reactions occurring at different temperatures. Such properties are central to the inherently safer design of processes involving highly exothermic reactions. In this work, calorimetric analyses are carried out for SCS under adiabatic conditions using Accelerating Rate Calorimeter (Phi-TEC I) in batch conditions. Such laboratory equipment provides highly accurate data acquisition of higher resolution temperature-pressure data versus time scale for accurate characterization of the rapid exothermic reaction. The maximum heat and pressure generation can be obtained and linked to the reaction enthalpy and the stoichiometric gas production. Here we study the reaction chemistry of nickel nitrate-fuel mixture in the Phi-TEC I for the synthesis of Ni nanoparticles at different fuel types, metal precursors, solvents, and amount of initial solution in a closed calorimetry cell. This study paves the way to identify the design requirements to scale up the solution combustion synthesis technology in a closed environment, which is an important step in scaling up this technology to an industrial production scale.

141a A Methodology for Commercial Process Technology Translations of Engineering Thermoplastic Resins

Rahul Patil

Introduction of novel product offerings into the marketplace is a complex act, especially when the key ingredient in the product formulation is a new molecule or new polymer. Engineering thermoplastic resins are no exception to these challenges. The journey from discovery of a new polymer to its successful commercial scale production cannot be taken for granted. In other words, discovery of a new polymer does not automatically make it a game-changer for the marketplace. Turning this new discovery into a viable business solution requires a methodical approach.

In this paper, the author will present one such methodology to convert the discovery of new polymer into a successfully commercialized resin production technology within existing assets. Consideration for the scale-up of any new polymer into an existing asset must start in the early phases of the research. The criteria for commercial scale production using existing assets must be clearly defined upfront. Then, the map of asset capability needs should be prepared during the bench-scale development. This early-stage adoption of mapping asset capabilities allows for the critical time needed to identify and study the fundamental principles applied to the production process. This is also a phase that requires assessment of an initial specification matrix needed for the new polymer that directly correlates to the Critical-To-Quality parameters of the product formulation. The next step makes a systematic use of pilot-scale process development using experimental design principles. Detailed risk assessment of the proposed commercial process footprint in the existing asset is important to identify piloting needs. Piloting allows for careful determination of the operating regime and potential failure modes across the entire production process. Lastly, commercial scale production tests are to be designed by a process technology team in close collaboration with all the different functions that are responsible for long-term term execution of a commercial production. The author will conclude by sharing recent commercialization examples that have followed this methodology with success.

141b In Situ Process Monitoring Improves Process Development

Michael DePierro

In situ reaction monitoring improves process development efficiency while reducing safety and quality risks. Process analytical tools enable gathering of real-time continuous data, providing insights into reaction behavior not possible with a one sample at a time approach. This presentation will discuss examples in which in situ monitoring has been used to expedite process development and improve efficiency. We will also highlight how process analytical tools have assisted in developing safe processes and minimizing safety risk during scale up.

55a Solute-Solvent NRTL: A Novel Approach to Model Amino Acid Solubility

Pedro Mallet Paula, Chau-Chyun Chen

Solution thermodynamic models are extensively used to describe fluid phase equilibria and thermodynamic properties for liquid mixtures. The Nonrandom Two-liquid (NRTL) activity coefficient model proposed by Renon and Prausnitz (1968) has been one of the most practiced solution thermodynamic models accurately describing the solution nonideality in liquid mixtures. However, the requirement of two binary interaction parameters per binary system with the NRTL model proves to be excessive when applied to multicomponent solute-solvent systems such as amino acid solubilities in various pure and mixed-solvents. This work presents a simplified NRTL formulation that reduces the number of binary interaction parameters from two to one for solute-solvent binary pairs and applies it to describe the fluid phase equilibria of different amino acid and solvent solutions. The model gives good predictions for the solubility of glycine, DL-alanine, DL-methionine, aminobutyric acid, isoleucine, leucine, norleucine, phenylalanine, serine, tryptophan, tyrosine, valine, diglycine, N-succinyl-L-phenylalanine-p-nitroanilide (SPNA), and L-phenylalanine-p-nitroanilide (HPNA) in various solvents. Furthermore, the simplified NRTL formulation can be employed in wide varieties of solution systems involving solutes, such as aqueous electrolyte solutions, CO_2 solubility, and membrane sorption, showing a critical potential for the application of the new formulation

55b Digital Materials Engineering for Development of Materials

Ravindra Aglave, Jan-Willem Handgraaf, Flavio Souza

Large technological advancements are almost always dependent on the creation (or discovery) of novel materials. Revolutionary developments in material science enable the next generation of technological advancement. The full exploitation of those technologies (by engineers) will often require further investment into the evolutionary development of those materials. Advanced materials with novel properties hold the key to innovation for organizations across industries. As the transition to clean energy gathers pace, there's increased pressure on design engineers and researchers to develop innovative materials for sustainable energy solutions.

However, the path from initial discovery of new materials to production can be costly and fraught with many challenges. This paper examines how simulation can be an effective way for organizations to discover and develop new materials faster to solve their complex engineering challenges.

In this talk we will speak about the entire value chain of materials: from discovery to performance assessment to production processes as well as post-production processing. We will also identify the current gaps in this workflow as well as our activities to close those gaps. The Siemens solution consists of an end-to-end workflow based on the digital twin concept with four main steps: materials discovery, materials performance analysis, process design and engineering, and materials processing. The talk will include examples of these phases to weave a digital thread through the solution as well as the tools that are involved both for simulation as well as the backbone (material lifecycle management) platform for digital engineering.

References

1. The new science of digital materials
2. Accelerate materials discovery & development
3. Composite materials state of the market

42b Modelling and Optimization of Bitumen Partial Upgrading Process Using HYSYS

Hung Hai Pham, Nam Sun Nho, Sukhyun Lim, Kwang Ho Kim, Sun-A Choi, Youngdoo Kim, Eun-Hee Kwon, Ki Duk Kim

The partially upgrading process is known as the process to transform bitumen just enough to reduce diluent requirements, lowering transportation costs and improving profit. In general, the partially upgraded process consists of a solvent deasphalting step to reject asphaltenes from bitumen, a thermal cracking of the deasphalted product step, and a hydrotreating step to saturate the olefins generated by thermal cracking. However, the process has been still observed to have difficulties in achieving the product quality targets set for partial upgrading (density <0.9400 g/cm^3 at 15.6 ^{o}C and viscosity <350 cSt at 7 ^{o}C). The process configuration is on of main factors to allow higher thermal conversion levels and product quality. Thus, there is a motivation to develop the partial upgrading process to overcome the above difficulties. In this study, a plant-wide process model was developed by considering visbreaking and solvent deasphalting units for investigating configurations of partial upgrading processes. The visbreaking and deasphalting unit models were built using the experimental data from the bench-scale facilities and from the literature. The process evaluation was done in terms of process performance and the equipment capex. Finally, an optimal process scheme was selected, which has shown the best potential to achieve the product quality targets

134a Better Together: Education and Collaboration to Mitigate the Risk of High Energy Materials in Research and Development Laboratories

Victor Sussman, Katie Mulligan

That hazard recognition is a critical antecedent to hazard *mitigation* will resonate with Process Safety and Environmental, Health, & Safety (EH&S) professionals alike. The research and development (R&D) setting presents several challenges to attaining high levels of hazard recognition, particularly with respect to high energy (HE) materials, that is, chemical compounds or mixtures that, through commonly encountered laboratory operations (friction, impact, heat, etc.) may undergo rapid and unpredictable decomposition with the concomitant release of significant stored chemical energy. As research activities frequently entail combining reagents in new and (hitherto) unanticipated ways in the pursuit of new knowledge, products, and technologies, the generation of HE materials can be an unintentional and undesired byproduct of experimentation.

The smaller scale of laboratory operations can inadvertently create a false sense of security that such activities are inherently safe. While broader community impacts are less likely, the potential for damage to equipment and potentially serious injury to personnel remain significant.

We will describe approaches to raising overall hazard awareness within R&D laboratories at The Dow Chemical Company. Our solution entails a three-pronged approach of basic education, decision-aids for researchers to help them assess potential hazards on their own, and a strong and effective partnership between individual researchers and reactive chemistry subject matter experts to provide timely feedback on proposed research activities with the goal of facilitating safe and effective experimentation in the laboratory. We will highlight several examples from industry and academia where insufficient hazard recognition resulted in reactive chemical incidents and include success stories from Dow laboratories where collaboration between researchers and reactive chemistry subject matter experts has resulted in safer outcomes and demonstrated that safety programs are enabling to the research enterprise.

134b Steam Cracker Thermal Imaging and Automated TMT Monitoring Using NIR Camera

Thomas Heinzerling, James Cross

Thermal imaging devices are commonly used for periodic inspection and measurement on steam crackers, but
less so for continuous thermal imaging and temperature data collection and analysis. Continuous monitoring
and repeatable, reliable measurements allow operators to extend furnace runtime closer to the material limits.
Automated detection of hot spots, together with temperature-controlled decoking, avoids accidental overheating
of the coils, increasing their lifetime and reducing maintenance costs.
A pilot installation in Gelsenkirchen, Germany highlights some key advantages and considerations when
implementing the technology. With real-time remaining coil lifetime prediction algorithms, software can predict
the risk of coil rupture before it becomes a substantial risk, giving operators the insights they need to optimize
replacement intervals while mitigating the risk of rupture,

134c Factors Impacting Level of Carbon Oxides from Use of Anti-coking Catalyst for Steam Cracking of Hydrocarbons (YieldUp*)

Nimeshkumar Patel, Joop Dees, Alain Pothuaud

Build-up of coke deposits in ethylene cracker furnace tubes is a problem that affects the reliability and productivity of ethylene plants worldwide. These deposits cause flow restrictions and pressure differential through the tubes, leading to maintenance downtime, shorter run-lengths and lost production. Veolia Water Technologies & Solutions (VWTS) has developed and patented the YieldUp anti-coking catalyst technology to assist ethylene producers in reducing the impacts of coke build-up, increasing furnace run-length two- to three-fold, as successfully demonstrated from full-scale field experience. This technology also offers potential to achieve higher conversion due to further improvement in anti-coking tendency at higher temperatures. YieldUp anti-coking catalyst works by gasifying coke upon contact, leading to the generation of higher levels of carbon oxides. This paper discusses the differences in level of carbon oxides generated from typical lab jet-stirred reactor, pilot-scale furnace, full-scale furnace computer simulation, and full-scale field trial.

*Trademark of Veolia; may be registered in one or more countries.

72a Properties of Cement Pastes Based on a Comparative Study of Various Rheological Models

Oluwaseyi Ayeni, Katherine Slamen, Bahman Ghorashi

Cement pastes are used in a range of industrial applications. In addition to construction industry, there are other applications of cement pastes, such as their use in petroleum industry for isolating high pressure zones and maintaining well integrity. In recent years, 3D printing has been employed in housing construction using various novel composite materials comprising cement and admixtures. Rheology which is the study of deformation and flow of matter can predict the behavior of cement pastes in processes, such as 3D printing.

Under shear, cement paste may behave as a non-Newtonian fluid. Various rheological models are reported in the literature, describing the viscosity dependence of cement on the rate of shear. In this study, seven rheological models were investigated to determine the range of their applicability, using a set of data for Portland cement. The rheological models used are Bingham Plastic, Power law, Herschel-Bulkley, Casson, Ellis, Carreau-Yasuda, and vom Berg. The results show which models are best fitted to low, mid-range, and high shear rates. Additionally, the inherent differences among a selected number of models, including their limitations, are noted, and evaluated. The evaluation of each model was based on a linear regression analysis of a set of experimental cement shear rate-shear stress data.

72b Simulation Study on Gas-Solid Separation Performance of Axial Flow Cyclone

Feng Chen, Haopeng Ding, Hao Yang, Zhihua Wang, Zhongli Ji

Cyclone separators are widely used in oil processing, natural gas purification, catalyst recovery and other fields. After gas-solid or gas-liquid two-phase fluid passes through the separator, discrete phase (particles or droplets) will be brought to the inner wall of the separator under the action of centrifugal force, and then sedimentation separation will occur. Generally, the pressure drop and efficiency of the cyclone increase or decrease at the same time. Therefore, in order to improve the separation efficiency of the axial flow cyclone under the condition of low pressure drop, a new blade projection parameter was designed in this work as the basic model for experimental verification and simulation analysis. The basic separation performance data of the cyclone at different flow rates were obtained by establishing a gas-solid separation performance evaluation system for the cyclone. The pressure distribution, velocity distribution and wall wear in the cyclone were studied by using computational fluid dynamics simulation method. The results showed that the designed cyclone blade could produce a limited swirl field, and had lower pressure drop and higher separation efficiency (more than 99%) at rated flow rate. Particles with diameter size larger than 5 μm could be completely removed using this cyclone. When the flow rate reached above the rated flow, the internal wear of the cyclone started to appear, and the wear position and wear amount basically did not change with the increase of the flow rate.

72c Thermodynamic Modeling of Adsorption Azeotropes

Nooram Anjum, Usman Hamid, Chau-Chyun Chen

Adsorption is a surface process that leads to transfer of a molecule from a bulk fluid to solid surface. Recently, adsorption-based processes have been expanding their presence in many industrial applications, due to their high selectivity, high throughput, and high energy efficiency, coupled with low maintenance and ease of operation. Particularly, they play a fundamental role in catalytic reaction processes and in separations and purification of gases and liquids [1]. An azeotrope occurs when the gas and liquid phase mole fractions in a binary gas mixture are equal at a given temperature and pressure, that is, when they intersect the 45-degree diagonal on an x-y diagram and the selectivity becomes unity. This azeotrope formation phenomenon is commonly found in vapor-liquid equilibria and gas-solid adsorption. Numerous articles have appeared in the literature reporting on the experimental formation of adsorption azeotropes in binary gas mixtures for a variety of adsorbate–adsorbent systems [2]. However, the complexities of solid surfaces and our inability to characterize exactly their interaction with adsorbate molecules limit our understanding of adsorption azeotropes.

To model such behaviors, modified one-parameter Margules equation with an exponential dependency upon the reduced grand potential and similarly modified van Laar-like equation are commonly used to express the adsorbate activity coefficients representing the adsorbed phase non-ideality leading to the formation of azeotropes [3]. Since these equations are purely empirical and they require the calculation of hypothetical reduced grand potential, a clear understanding of this azeotrope formation phenomenon has not been fully developed so far. The study presents thermodynamic modeling of adsorption azeotropes based on the rigorous thermodynamic framework of adsorption Nonrandom Two-liquid (NRTL) [4] model for adsorbed phase coupled with generalized Langmuir model for multicomponent adsorption equilibria [5]. This comprehensive thermodynamic framework allows critical analysis for the formation of azeotrope in adsorption equilibrium at system temperature and pressure without the complication of the reduced grand potential.

References:

1. Jain, S., Moharir, A.A., Li, P., Wozny, G.: Heuristic design of pressure swing adsorption: a preliminary study. *Purif. Technol.* 33, 25–43 (2003).
2. Reeds, J. N.; Kammermeyer, K. Adsorption of Mixed Vapors. *Eng. Chem.* 1959, 51, 707–709.
3. Luberti, M., Mennitto, R., Brandani, S., Santori, G. and Sarkisov, L., 2021. Activity coefficient models for accurate prediction of adsorption azeotropes. *Adsorption*, *27*(8), pp.1191-1206.
4. Kaur, H., Tun, H., Sees, M. and Chen, C.C., 2019. Local composition activity coefficient model for mixed-gas adsorption equilibria. *Adsorption*, *25*(5), pp.951-964.
5. Hamid, U., Vyawahare, P., Tun, H. and Chen, C.C., 2022. Generalization of thermodynamic Langmuir isotherm for mixed-gas adsorption equilibria. *AIChE Journal*, *68*(6), p.e17663.

Role of Calcium ion in bio-organic fertilizer production and diffusion kinetics into the soil for plant growth

Rhyin Mandal [1], Anindita Sen [1], Rwiddhi Sarkhel [2], Tamal Mandal [2]*

[1] *Hem Sheela Model school, Durgapur, India*

[2] *Department of Chemical Engineering, National Institute of Technology (NIT) Durgapur, Mahatma Gandhi Avenue, A-Zone, Durgapur, West Bengal, India, 713209.*

[2*] *Corresponding author: Prof.Tamal Mandal, Department of Chemical Engineering, National Institute of Technology (NIT) Durgapur, Mahatma Gandhi Avenue, A Zone, Durgapur, West Bengal, India, 713209.*

Email: tamal.mandal@che.nitdgp.ac.in

ABSTRACT

Waste valorization or value recovery is the valorization of waste products by reuse or recycling in order to create economically useful materials i.e. producing value added products from waste, converting waste to value. This study mainly deals with the purpose of bio-fertilization and the transportation of nutrients to the soil to reach the plant roots. Biofertilizers play an essential role in colonizing the rhizosphere and supply of NPK for promoting growth to the host plants. Continuous application of chemical fertilizers exhibited detrimental effects on the environment and Agroecosystems. Therefore, the development of bio-organic fertilizers has been a solution to the problem. This study deals with the soil diffusion kinetics and N, P, and K constituent distribution for the bio-fertilizer and the calcium to the soil. The mathematical models have been applied to determine the liberation of nutrient sources for plant growth, transport of fertilizer components in the soil, and absorption in plant roots for better productivity by soil diffusion kinetics. The dissolution rate of the applied fertilizer depends upon the temperature, surface area, pH and ionic groups present in the molecule. Therefore, the rate of dissolution was based on equation written as $-\frac{dc}{dt} = kC^n$. The rate constant and order of dissociation of fertilizer was evaluated using kinetic study by plotting $ln\frac{\Delta C}{\Delta t}$ vs. lnC. The concentration profile of the bio-organic fertilizer can be calculated by the equation:

$$C_A(y,t) = \int_{y'=-\propto}^{y'=+\propto} \frac{C_A(y')}{\sqrt{4\pi 0.92\times10^{-2}\,t}}\, e^{-\left[\frac{(y-y')^2}{4\times0.92\times10^{-2}\,t}\right]} dy'.$$

For bio-organic fertilizer $C_A = e^{-kt}\int_{-\propto}^{+\propto} \frac{1}{\sqrt{4\pi 0.92\times10^{-2}t}}\, e^{-\left[\frac{(y-y')^2}{4\times0.92\times10^{-2}t}\right]} C_A(y')dy'$

The solution for the diffusion kinetics as represented in the above equation indicates the fertilizer concentration with respect to time and positions. It is considered that adsorbed ingredients of fertilizer are converted chemically and/or bio-chemically of the plants following first-order reaction kinetics. The diffusibility of the nutrients to the soil by mass transport. So, the no of molecules of water transported to the soil is represented as:

$$N_{H_2O} = D_{H_2O\,eva} \frac{dC_{H_2O}}{dz} + D_{H_2O\,PI} \frac{dC_{H_2O}}{dz}$$

$$= \left(D_{H_2O\,eva} + D_{H_2O\,PI}\right) \frac{dC_{H_2O}}{dz}$$

$$= D_{H_2O\,eva\,PI} \frac{dC_{H_2O}}{dz}$$

The flux at any time and position for chemical and bio-organic fertilizers are

$J_{Ay} = \frac{5\sqrt{0.495}}{\sqrt{4\pi t}} e^{-\left[\frac{y^2}{4\times 0.495t}\right]}$ (chemical fertilizer) and $J_{Ay} = \frac{5\sqrt{0.92\times10^{-2}}}{\sqrt{4\pi t}} e^{-\left[\frac{y^2}{4\times 0.92\times10^{-2}t}\right]}$ (bio-organic fertilizer). This shows that the developed biofertilizers as compared to the chemical fertilizers show better results in terms of dissolution rate and absorption capacity.

***Keywords**: Waste valorization; sustainable; staple; mathematical model; fertilizers.*

GRAPHICAL ABSTRACT

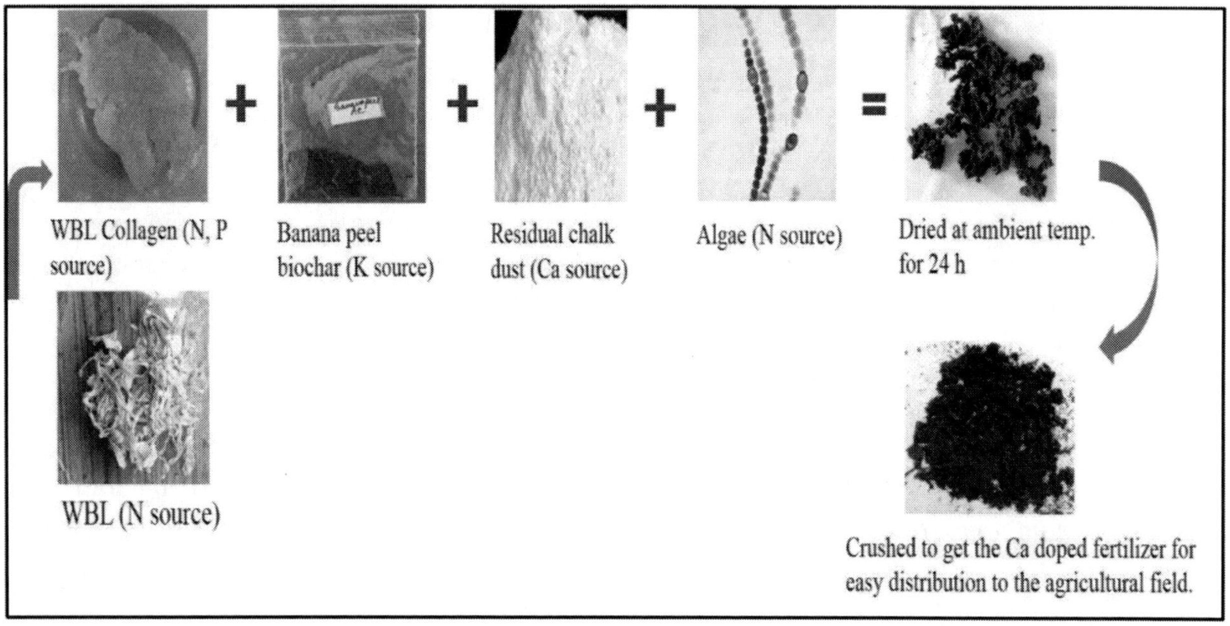

1.0 INTRODUCTION

Disposal of residual wastes nowadays has been a great problem. Solid wastes comprise disposal through unsanitary practices and piling up around cities and different countries because of economic problems, lack of technologies and facilities (Patel et al., 2014; Saedi et al., 2016). Processes of collection, transmission and landfill are very expensive and threatening public health and the environment (Saedi et al., 2016). So, developing environmentally feasible methods of waste utilization is of utmost importance. Two important issues have been addressed for the present study: waste management and soil fertility (Nogueira et al., 2011).

Enhancing the fertility of soil increases the crop production so followed as the dependence on inorganic chemical-based fertilizers, cause a serious threat to human health and the environment. Excessive use of commercial inorganic fertilizer poses a detrimental effect on environment and Agro-ecosystems, decreases the organic matter content; decline soil pH; and finally deteriorate soil health (Lin et al., 2019; Abera et al., 2018; Li et al., 2012). As it is known that fruits and vegetables grown artificially threatens human life and grown on inorganically over-fertilized soil are more prone to pests and diseases (Karungi et al., 2006; Parmar and Chakraborty et al., 2016). The objective of biofertilizer is to increase the soil fertility, efficiency of the soil and reduce the use of inorganic fertilizer to establish continual Agro ecosystem. Therefore, the solution to defeat the problem is the development and utilization of a bio-organic fertilizer for better crop production and high yield of fertility. The literature studies have shown that organic fertilizer as compared to inorganic based fertilizer has higher total nitrogen (N_2) and organic matter content, which can help to improve the soil's physical, chemical, and biological property by increasing the population of useful microbes in the soil (Alvarez et al., 1988). This fertilizer contains the nutrients needed for plant growth in absorbable form such as nitrate, phosphate, exchangeable calcium, dissolved potassium and other substances; on the other hand, it has a wide contact surface suitable for microbial activity, and thus preparing a variety of nutrients for consumption (Gao et al., 2019). Fertilizer is the main component that add nutrients to the soil and promoting soil fertility and plant growth. Chemical NPK fertilizers are used by the farmers to enhance agricultural yield and soil fertility, plant growth. Continuous use of chemical fertilizers damage soil characteristics and reduces soil fertility (Kochakinezhad et al., 2012).

Biofertilizers are those that produces supplementary substances needed for the growth, development and productivity of plants. These fertilizers contain live and efficient formulates of bacteria, fungi or blue-green algae either separately or in combination those can fix atmospheric nitrogen, solubilizes phosphorus and decomposes organic matter (Singh et al., 2022). Enormous amount of waste such as chrome tanned wet blue leather (WBL) is generated by leather industry which causes a toxic effect on the environment and human health. To prevent leather from microbial destruction, the leather industry typically uses tanning agents such as chromium-containing salts. Chromium causes a harmful effect and negative consequences on the environment since it is carcinogenic and a toxic heavy metal (Tasca and Puccini et al., 2019; Kolomaznik et al.,2008; Dhal et al., 2013). When it is inhaled, they can enter the respiratory tracts causing pulmonary sensitization. Also, it causes lung and nasal ulcers and cancer (Mertz et al., 1969), as well as bronchial asthma (Park et al., 1994), skin allergies (Adams et al., 2006), and reproductive and developmental disorders (Langerd et al., 1990; Shekhawat et al., 2015). So, development of bio-based fertilizer has been a solution by formulating a nitrogen source from WBL through the pretreatment process by acid hydrolysis removing chromium salts from WBL. But after the removal of chromium from WBL, amount of chromium left is about 22.4 % almost negligible trace amount and is acceptable for soil by WHO, 1996 (World Health Organization).

In the present study, development of liquid fertilizer has been focused as it is more effective and convenient to use in agricultural productivity. Liquid biofertilizers are used as suspensions with microbial assays, which may fix atmospheric nitrogen and can solubilize insoluble phosphates to increase the productivity of crops and enhancing the fertility of soil (Mitter et al., 2021). Bioorganic liquid fertilizer not only increases bioorganic fertility of crops (in comparison to the control and prototype fertilizer), but also accelerates their maturation and nutrient quality. Liquid biofertilizer can be used as an inoculant for the growth of many plants and vegetables. Bio-organic liquid fertilizer has been planned to develop combining the N, P, and K source using leather waste and agricultural waste such as banana peels. Waste valorization, value recovery is the process of waste products or residues from an economic process being valorized, by reuse or recycling in order to create economically useful materials. The term comes from practices in sustainable waste management (Kabongo et al., 2013).

Biofertilizers play an essential role in colonizing rhizosphere and promoting growth to the host plants. Continuous application of chemical fertilizers exhibited detrimental effect on the environment and Agro-ecosystems. Development of bio-organic fertilizers has been a solution to the problem. This study deals with the soil diffusion kinetics and N, P, K constituent distribution for the biofertilizer.
Banana, being a highly consumed fruit in tropical countries of the world. It is easily available and cost-effective. After harvest, almost 60% of banana biomass is left as waste. Worldwide about 114.08 million metric tons of banana peel waste loss are produced, leading to environmental problems such as excessive emission of greenhouse gases. Banana peels are utilized as it is highly consumable in tropic countries and easily available. These wastes contain high content of paramount industrial importance such as cellulose, hemicellulose, and natural fibers that different processes can modify such as bacterial fermentation and anaerobic degradation (Acevedo et al., 2021).
Calcium doped biofertilizer has been developed by combining residual waste chalk dust and biofertilizer. Residual waste chalk dust was obtained from teaching classrooms. Chalk, due to its physical properties and high reactivity, seems to be an interesting component for production of various types of fertilizers. The paper presents comprehensive research on chalk based fertilizing mixtures. Maintaining the proper soil pH is one of the most important factors influencing the productivity of Agro-ecosystems. The soil reaction influences the sorption complex capacity, microbiological activity and the effectiveness of the use of fertilizer ingredients. The focus is concerned with the addition of calcium carbonate through chalk dust in the soil and how it helps in increasing the nutrient content of the soil. Calcium ions are responsible for holding together the cell walls of plants. Calcium carbonate is basic in nature so reduces the pH level ensuring a healthy growth of plants. Calcium can be used on a large scale as fertilizer to increase the yield of plants without artificial fertilizers such as ammonium nitrate, ammonium phosphate, superphosphate, etc. (Mudryk et al., 2018).
In this study, mathematical models have been applied on liberation of fertilizer components from their source, transport of fertilizer components in the soil, and absorption in plant roots which indicates the optimum fertilizer use for better productivity. It has been tried to follow the liberation of plant nutrients and water from fertilizers to plants. The three important sections have been identified to play the active role in this whole process; Firstly, the liberation of nutrient components from fertilizer to soil-water; Secondly; Transportation of plant nutrients through the soil-water system by diffusion mechanism and finally, absorption of plant nutrients in plant roots by diffusion mechanism (Majee et al., 2023).
It is considered that adsorbed ingredients of fertilizer are converted chemically and/or bio-chemically of the plants following first-order reaction kinetics, the partial differential equation chosen being the system becomes:

$$\frac{\partial C_{Ai}}{\partial t} = D\frac{\partial^2 C_A}{\partial y^2} - K_A\, C_A \qquad (1)$$

The solution to the above equation is obtained by first considering that no diffusion occurred, the differential equation is given by $\frac{dC_{Ai}}{dt} = -K_A\, C_{Ai}$ (2)

With solution $C_{Ai} = C_A(y,t) = C_{Ai0}(y)e^{-kt}$ (3)

$C_{Ai0}(y')$ = initial concentration of any component at any position y in system.

This suggests the following solution to equation (9), Let

$C_A(y,t) = C_{A0}\,(y,t)e^{-kt}$, therefore, $C_A = C_{A0}$ at $t = 0$ (4)

Substituting equation (4) into equation (1) gives $\frac{\partial C_A}{\partial t} = D\frac{\partial^2 C_A}{\partial y^2}$ (5)

The solution of equation (1) is obtained from equation

$$C_A(y,t) = \int_{y'=-\propto}^{y'=+\propto} \frac{C_A(y')}{\sqrt{4\pi D_{AB}t}} e^{-\left[\frac{(y-y')^2}{4Dt}\right]} dy'$$

So, that the ultimate solution $C_A = e^{-kt} \int_{-\propto}^{+\propto} \frac{1}{\sqrt{4\pi Dt}} e^{-\left[\frac{(y-y')^2}{4Dt}\right]} C_A(y')dy'$ (6)

For chemical fertilizer $C_A = e^{-kt} \int_{-\propto}^{+\propto} \frac{1}{\sqrt{4\pi 0.495t}} e^{-\left[\frac{(y-y')^2}{4\times 0.495t}\right]} C_A(y')dy'$ (7)

For bio-organic fertilizer $C_A = e^{-kt} \int_{-\propto}^{+\propto} \frac{1}{\sqrt{4\pi 0.92\times 10^{-2}t}} e^{-\left[\frac{(y-y')^2}{4\times 0.92\times 10^{-2}t}\right]} C_A(y')dy'$ (8)

This indicates that concentrations in different positions with respect to time varies for the fertilizer components and the surface of the root of plants may indicate the water retention capability and absorption by the roots. Amount of fertilizer can be calculated for proper use by plants.

2.0 MATERIALS AND METHODOLOGY:

2.1 Materials:

WBL for nitrogen source was obtained from Leder Chemische Pvt, Ltd situated in Tangra, Kolkata, West Bengal. Residual chalk dust was collected from the teaching classrooms from Department of Chemical Engineering, NIT Durgapur, Paschim Bardhaman, West Bengal for calcium source. Banana peels was obtained from hostel mess NIT Durgapur for the source of potassium. Chemicals such as orthophosphoric acid (85%), hydrochloric acid (35%), and sodium hydroxide pellets were utilized for the experimental study. Double distilled water was used throughout the study. The formulated NPK calcium doped bio-organic has been evaluated by different characterizations: FTIR, XRD and SEM analysis from COE, NIT Durgapur.

2.2 Methods:

2.2.1 Procuring nitrogenous collagen source from chromium free WBL

Chromium free Wet blue leather (WBL) was extracted following three steps of acid pretreatment method using orthophosphoric acid by the method as developed in (Majee et al., 2021b). The chromium removal procedure was performed three consecutive times. First step, 6 g leather was mixed with 300 mL orthophosphoric acid of pH 1 and kept in a BOD incubator cum shaker maintaining a temperature of about 35 °C for 24 hrs. Then, the obtained chromium free solution of WBL was then separated by filtering using Whatman 42 filter paper, and this method was repeated twice. The treated nitrogenous collagenic WBL residue was properly washed with distilled water twice. The mechanism of chromium extraction is explained in the equation represented below as follows (Lima de Oliveira et al., 2008):

$$WBL - Cr + H3PO4 \rightarrow Collagen + CrPO4 \quad (9)$$

Fig 1: Chromium free WBL

2.2.2 Preparation of carbonized banana peel

Banana peels is utilized for the study as it is highly available and highly consumable in tropical countries. Raw banana peels were collected, sun dried and cut into small pieces of 2 cm each. The obtained pieces of banana peels were carbonized in muffle furnace at 400°C for about 2 hrs. Thus, the obtained carbonized form of banana peels was used as a biofertilizer for good yield and the growth of plants since banana peels contain high amount of potassium of about 22-40.25% and phosphorus 3.25% (Zaini et al., 2022). Proximate analysis has also been performed in the simulated laboratory conditions utilizing banana peels as the raw material. It is used to calculate the approximation of moisture, ash, and volatile matter contents in the available raw material. This process serves as an inexpensive and environmentally feasible method.

Fig 2: Formation of carbonized banana peel powder

2.2.3 Preparation of chalk dust as a Ca^{2+} source

Chalk sticks was collected from teaching classrooms, was pulverized into powders. Then the pulverized form of chalk was dissolved in water and was decanted at least 3-4 times to separate the excess water. Finally, a clear transparent solution of chalk dust was obtained which has been utilized as a carbonate-based fertilizer to enhance the yield of plant growth and strengthening the roots and stem of plants.

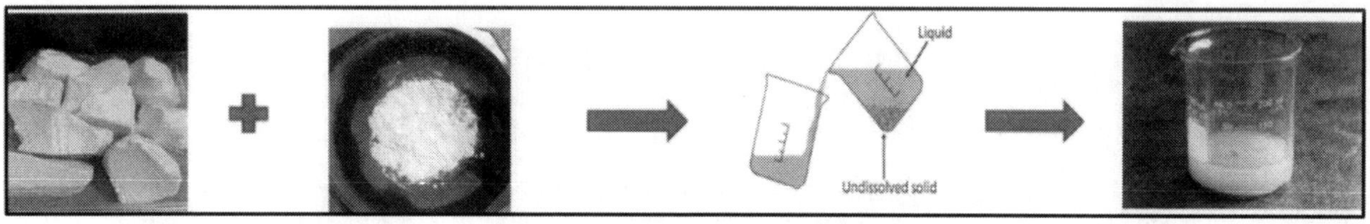

Fig 3: Decanted Chalk solution

2.2.4 Calcium doped NPK fertilizer production

The bioorganic NPK fertilizer was prepared by combining chromium separated collagenous ingredient of WBL (as a nitrogen source), banana peel biochar (as potassium source), and decanted chalk dust powder (as calcium source). These ingredients were uniformly mixed and dried at an ambient temperature of 30-35 °C for 24 hrs. The source of calcium carbonate was obtained from waste residual chalk dust from teaching classrooms. Finally, the hard manure was granulated into powder by mortar pestle to be utilized as a fertilizer for easy distribution to the agricultural field.

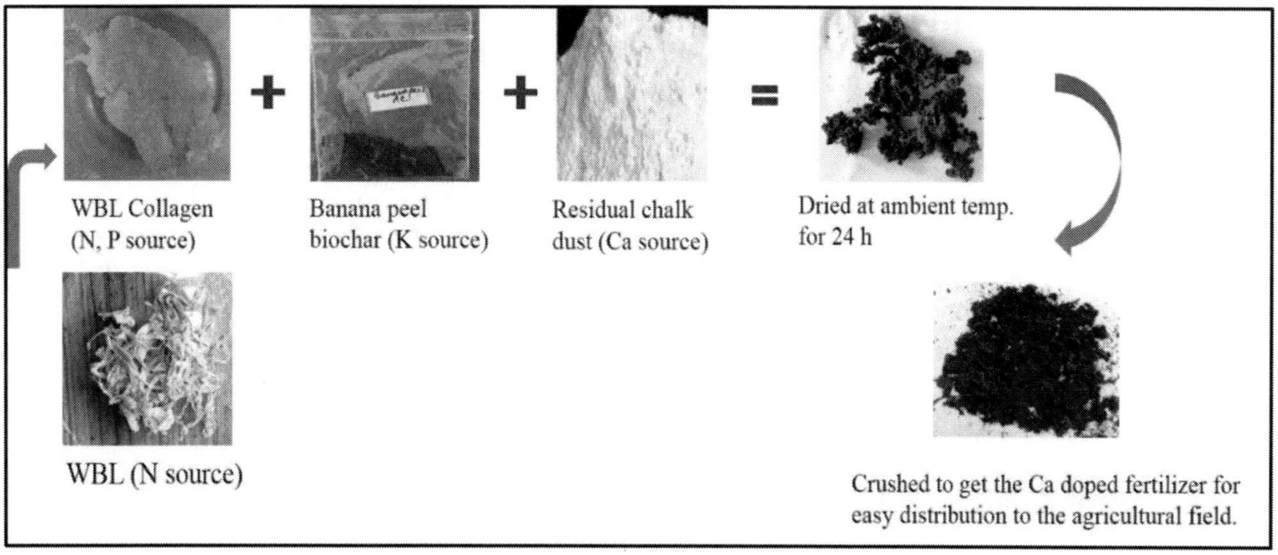

Fig 4: Calcium doped bioorganic fertilizer

2.2.5 Plant growth study on brinjal and chili using newly developed Ca doped NPK fertilizer

The plant growth study was interpreted on brinjal and chili plant. The size of leaf and stem was studied utilizing the calcium doped bio-organic liquid fertilizer. The fertilizer was sprayed on the plants for diffusion through soil. The calcium-based carbonate fertilizer was mixed with the soil for the growth of plants. The experiment was conducted with and without calcium doped fertilizer (control).

3.0 RESULTS AND DISCUSSIONS:

3.1 Characterization of WBL (SEM & EDX)

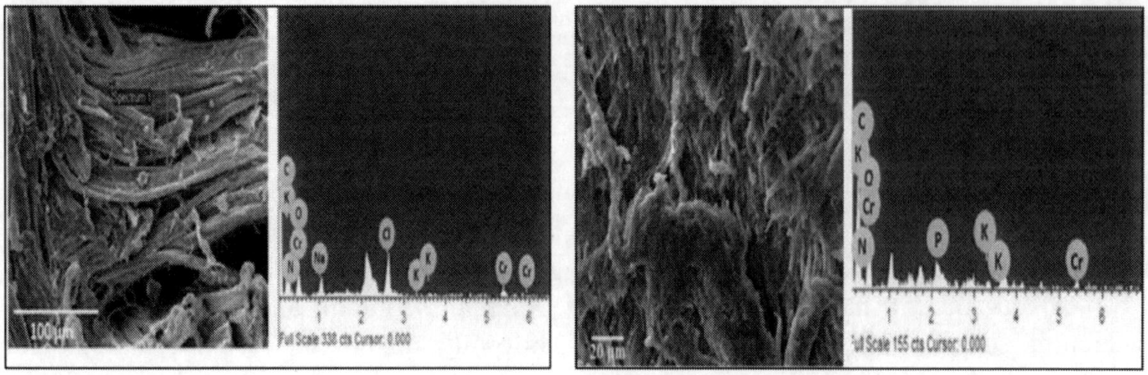

Fig 5 (a, b): SEM of Raw WBL and WBL after chromium removal

The surface morphology and topographical image of WBL was studied by SEM (EVO MA 15 / 18, CARL ZEISS) of COE, NIT Durgapur represented in Fig 5. Before the treatment, it showed the presence of collagen fiber structure stuck to each other with limited disorientation and looked like a thick bundle. 17.26 % nitrogen (N), 11.81 % chlorine (Cl), 39.13 % carbon (C), 0.23 % potassium (K), 6.55% chromium (Cr) (Majee et al., 2021).

After the treatment of chromium removed WBL, it has shrinked up collagen fibers which changes into an undifferentiated structure. About 90.92 % Chromium was removed from the WBL after three consecutive steps, and it reaches non detectable after six steps. Chromium was decreased by 0.46 %, indicating removal of chromium. Other elements have also been detected in EDX spectrum analysis 15.21 % nitrogen (N), 42.84 % carbon (C), 3.16 % phosphorus (P), 0.21 % potassium (K) (Majee et al., 2023).

3.2 Characterization of Ca doped NPK fertilizer

The X-ray powder diffraction study of $CaCO_3$ & Ca doped fertilizer samples was conducted with an X-ray diffractometer using Cu-Kα ($\lambda = 1.542$ Å) radiation at 50 kV and 1 mA. The XRD data were recorded in a 2θ range of 4–39° with an angular step width of 0.02° and a counting time of 1° (2θ)/min. XRD patterns of the residual chalk powder ($CaCO_3$) and the newly developed Ca doped NPK fertilizer has been shown in the fig below (Fig 6). $CaCO_3$ exhibited strong diffraction patterns with high range of crystallinity at angles of 10º, 22º, 28º, 32º with (012), (104), (110), and (202) reflections (Malinowski et al., 2007). The presence of calcium can be interpreted from 32.4º subjected to (202) planar reflections. But when the Ca is doped with the bioorganic NPK fertilizer, due to the porous nature of chalk revealing that they are semi-crystalline in structure and tend to be in an amorphous phase due to high energy levels at angles of 30º, 38º, 42º, 50º (Fig 6 b). In fig 6 b the presence of calcium ions is visible in 30º. It can be noticed that the crystallinity decreases enhancing the energy levels by reinforcing the residual chalk as a source of calcium into the formulated bioorganic NPK fertilizer. And the presence of potassium is maximum in the Ca doped NPK fertilizer in the form of potassium carbonate at an angle of 15.74º with (011) planar reflection (Lloyd et al., 1973). SEM images interprets the topographical changes on the surface of the particles. The morphology of decanted chalk powder was investigated by SEM (CARL ZEISS) and represented in Fig 6 (c, d). Before the treatment, it showed the presence of collagen fiber with disorientation of chalk sticks. EDS analysis represents 11.96 % Calcium (Ca), 48.2 % oxygen (O), 17 % carbon (C), 2.6 % potassium (K). The SEM images intimates the porous nature of chalk. The morphology of Ca doped NPK bio-organic fertilizer was investigated. After the treatment, it showed the presence of collagen fiber with reinforcement of potassium and calcium source into it. EDS analysis represents 15.1 % Calcium (Ca), 81.2 % oxygen (O), 3.1 % potassium (K). No presence of carbon has been noticed. The microporous cavities of calcium carbonate and banana peels has been estimated in the figure.

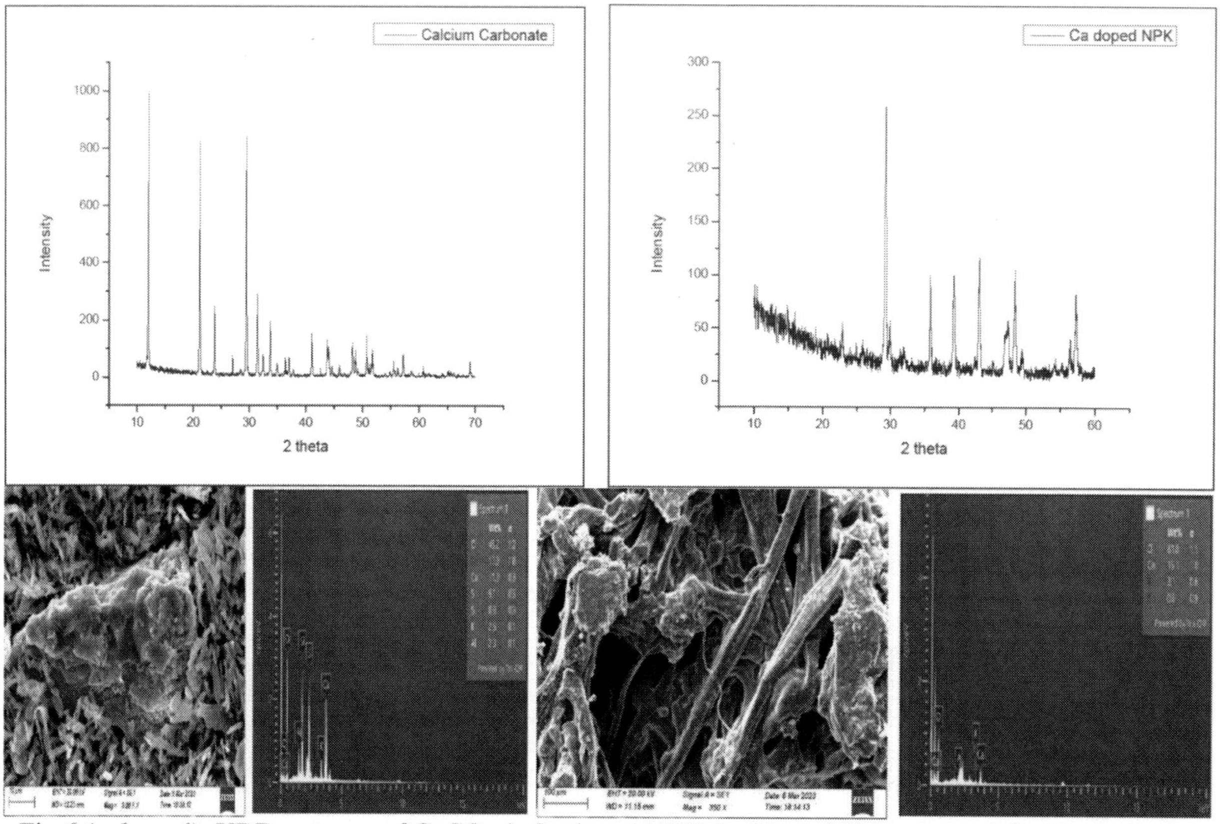

Fig 6 (a, b, c, d): XRD patterns of $CaCO_3$ & Ca doped NPK fertilizer; SEM images of $CaCO_3$ & Ca doped NPK fertilizer.

3.3 Optimum weight & pH analysis of different acid, alkali reacted with chromium free WBL

Different acids such as H_3PO_4, HCl, and alkali such as NaOH was treated with the chromium free WBL. Then it was analyzed at different pH. With increasing pH, the dissolving rate of WBL-Cr has been increased and optimum value is different for all acid/alkali as presented in Fig 7. It is 3 for HCl at 105 mg/l. For H_2SO_4, opt value is 2 at 800 mg/L. For NaOH, opt value is 9 at 920 mg/L. And for H_3PO_4, opt value is 1 at 800 mg/L (as shown in Fig 7). This dissolved material will be used as liquid biofertilizer after adding some more trace elements which is essentially required for crop production. In the present study, Calcium ions has been utilized from the waste residual chalk from the classroom. The optimized value for the weight of chromium free WBL at pH is shown in the figure (Fig 7) below.

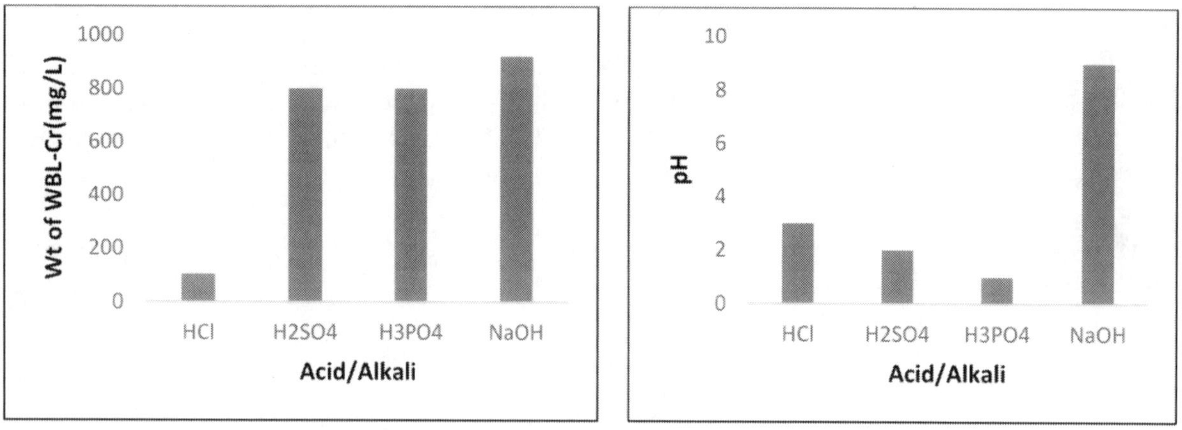

Fig 7: Optimum weight value of acid alkali at maximized pH

3.4 Characterization of banana peels

3.4.1 Proximate analysis of banana peels

Runs	Type of Banana peels	Moisture Content (MC)	Volatile Content (VC)	Ash Content (AC)	% Fixed Carbon (FC)
01.	Ripe cavendish banana	0.35	26.5	0.04	73.11
02.	Ripe cavendish banana	3.5	20.5	0.447	75.53
03.	Unripe Plantains	0.86	41.7	3.2	54.24
04.	Unripe Plantains	4.8	30.4	0.78	64.02

Formulae : $\% \boldsymbol{FC} = \mathbf{100} - (\mathbf{MC} + \mathbf{VC} + \mathbf{AC})$

On avg for ripe bananas, % Fixed Carbon (FC) = 74.3; On avg for ripe bananas % Fixed Carbon (FC) = 59.13; Total % Fixed Carbon (FC) = 66.7

3.4.2 Characterization of banana peels (SEM, XRD & FT-IR)

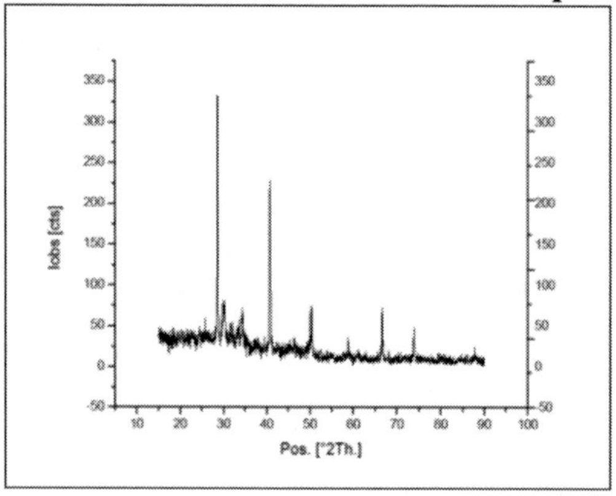

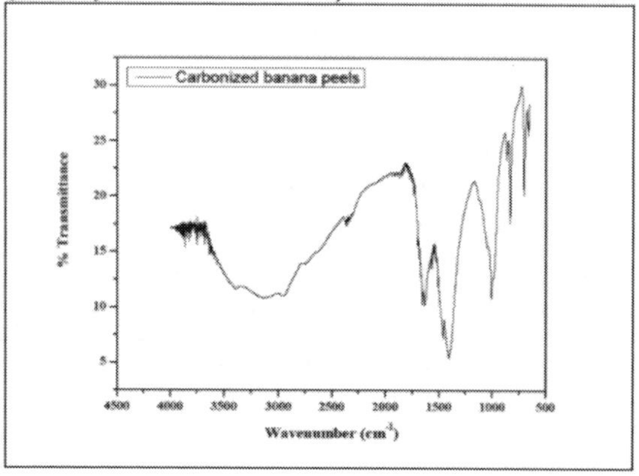

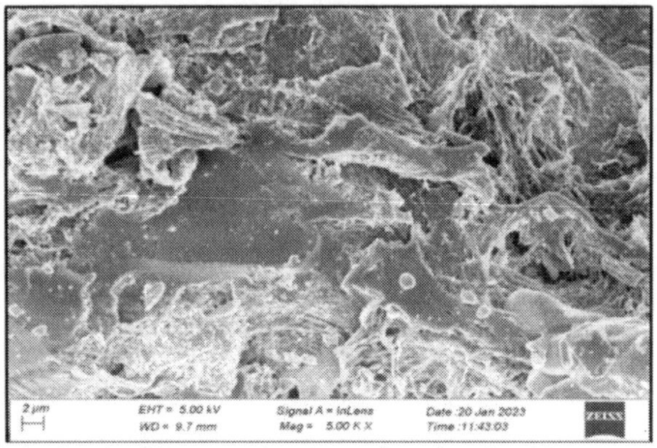

Fig 8 (a, b, c): Characterization of carbonized banana peels

XRD (X-ray diffraction) patterns of banana peels exhibited strong diffraction patterns revealing that they are semi-crystalline in structure and tend to be in an amorphous phase. The intensity of XRD peaks were shown due to the microporous nature of banana peels at maximum angles of 32°, 43° with planar reflections of (113) and (004) in the form of potassium phosphide.
SEM (Scanning electron microscopy) analysis of carbonized banana peels shows microporous structure and cavities on the surface of the banana peels as depicted in the figure above (Fig 8c). The banana peels show a high carbon and potassium content about 41% estimated by the EDAX analysis which interprets that the banana peel has been carbonized before use.
The FTIR infrared spectrum of the carbonized banana peels was used for the determination of functional groups with K Br pellets recorded in the wavelength range of 4500-500 cm^{-1}. A broad peak at 3640 cm^{-1} is due to the free hydroxyl group. The bonds around 2960 cm^{-1} denotes hydroxyl band of amino acid groups. The alkane C-H stretching bond is represented between 3000 cm^{-1} – 2800 cm^{-1}. The observed peak at around 1700 cm^{-1} is attributed to C=O stretching vibrations of carboxylic groups (-COOH, -$COOCH_3$) which can be due to the presence of carboxylic acids or their esters (Li et al., 2012). The carbon bond is visible in the -COOH bond due to the carbonization of banana peels and high activation energy (Fig 8 b).

3.5 Comparison of leaf and stem size of brinjal & chili plant

It has been observed from the graphs that there is an increase in leaf and stem size in case of bio-organic + calcium doped fertilizer as compared to control, chemical fertilizer and bio-organic fertilizer due to the formulation of NPK and amalgamation of calcium ions as depicted in Fig 9. It has been investigated that there is an increase in the length of stem and leaves since the chalk utilized as carbonate-based fertilizer has been used to strengthen the roots of plants which makes the soil beneficial for efficient diffusibility and the growth of plants. In case of Chili plant, a significant increase for all types of fertilizer is noticed due to soil characteristics in Durgapur, India. The soil is a little saline in nature whereas in case of brinjal, due to land degradation, soil characteristics is affected in Indonesia, Merjosari District, Lowokwaru Subdistrict, Malang which affects the productivity of brinjal.

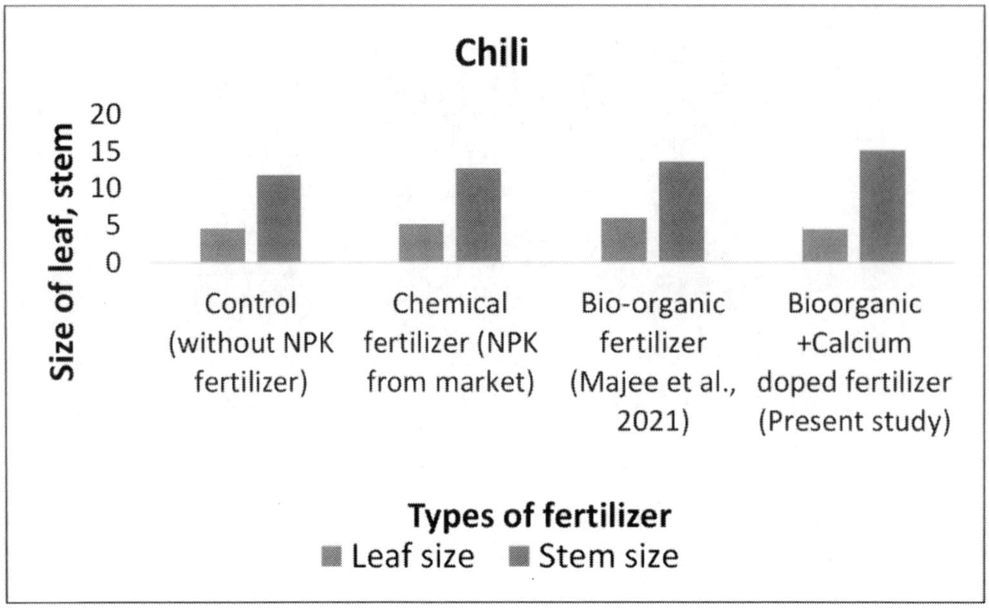

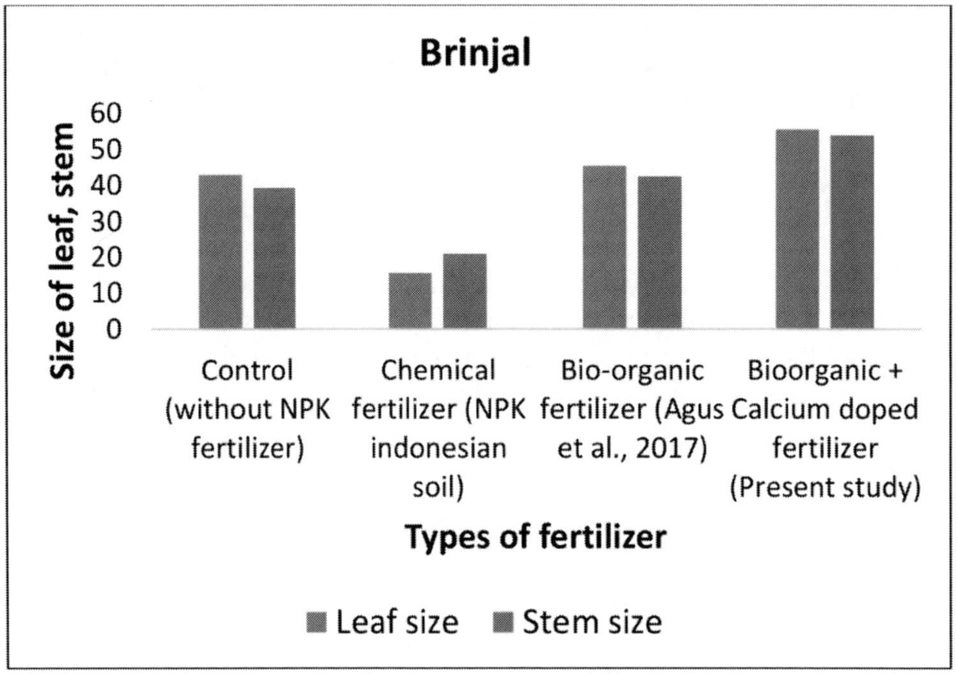

Fig 9 (a, b): Leaf and stem size for different fertilizer

3.6 Present study compared with relevant literature

The calcium ions have been utilized from waste residual chalk dust collected from teaching classrooms. The present study of bio-organic + calcium doped fertilizer has been compared with relevant literature of brinjal (Agus Suryanto et al., 2017) as well as chili plant (Majee et al., 2023). Brinjal shows a considerable increase in chemical fertilizer as compared to chili of about 5% whereas chili shows significant increase in calcium doped fertilizer of about 3%. Probably, this is due to the reason that chili plants can absorb the calcium ions in calcium doped bio-organic fertilizer more efficiently than Brinjal as shown in Fig 10 (Majee et al., 2021; Suryanto et al., 2017).

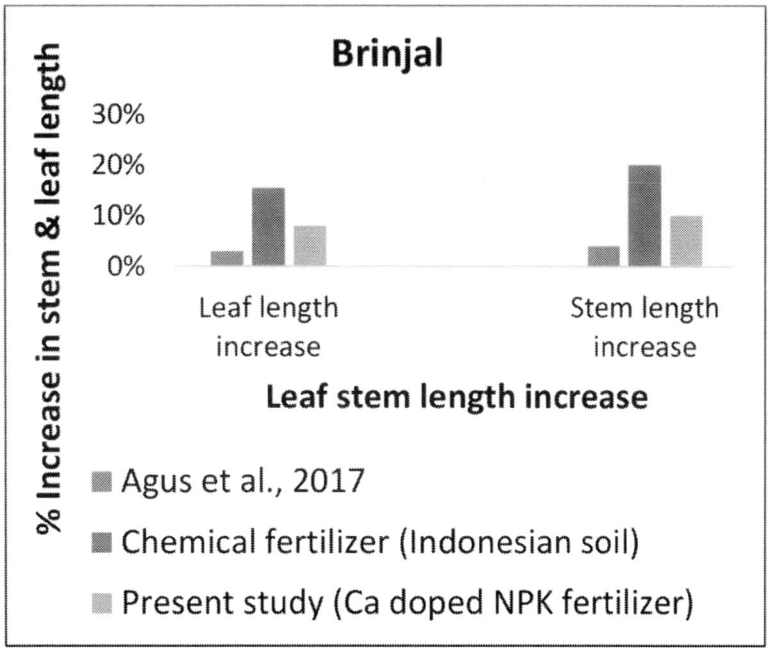

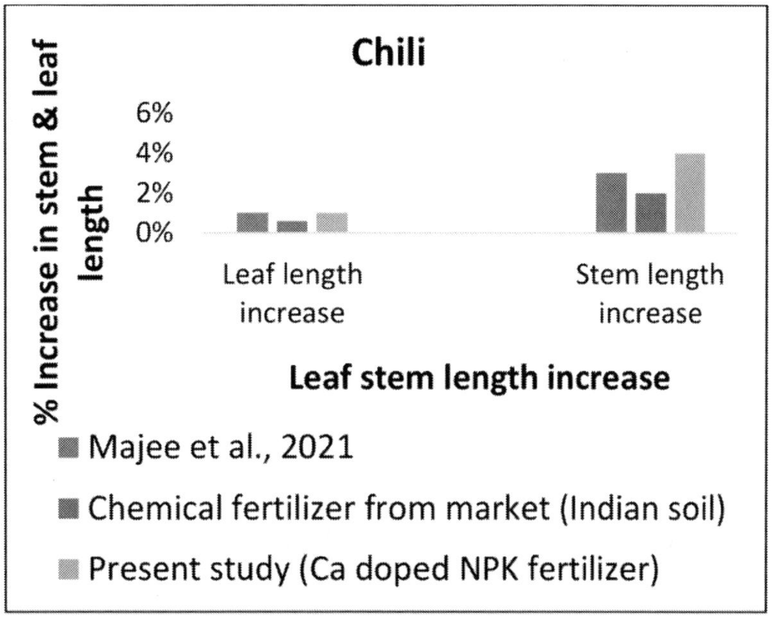

Fig 10: % increase in leaf, stem size on chili & brinjal plant using different fertilizers

3.7 Mathematical modelling

In this study, mathematical models have been applied on liberation of fertilizer components from their source, transport of fertilizer components in the soil, and absorption in plant roots which indicates the optimum fertilizer use for better productivity. It has been observed that the liberation of plant nutrients and water has taken place from fertilizers to plants. The three important sections have been identified to play the active role in this whole process; Firstly, the liberation of nutrient components from fertilizer to soil-water; Secondly; Transportation of plant nutrients through the soil-water system by diffusion mechanism and finally, absorption of plant nutrients in plant roots by diffusion mechanism (Majee et al., 2023).

Applied fertilizer was dispersed in the soil to identify the rate of dissolution. Usually, the dissolution rate depends on the surface area of the plant, temperature, pH, and functional ionic groups present in the fertilizers. It is considered that the plants are absorbing nutrients from the dissolve portion of nutrients. Therefore, the dissolution rate was considered on equilibrium concentration of nutrients in soil water mixture, and consequently, the governing equation can be written as $-\frac{dc}{dt} = kC^n$.

4.0 CONCLUSION

During the growth of different plants, the major factor responsible is the nutritional access in the soil. The focus of this study is waste valorization, i.e. converting waste products into value added products. The characterizations interpreted the NPK amalgamation with calcium doped fertilizer from waste sources like chalk which is rich in carbonate-based fertilizer. Calcium doped NPK fertilizer developed from collagenous material of WBL (N source), carbonized banana peels (K source), Chalk dust (Ca source), and orthophosphoric acid (P source). It can be concluded that this fertilizer enhances the crop yield and soil fertility. Also, the efficacy of formulated Ca doped fertilizer is higher compared to lab research works and other literatures.

Model development of soil diffusion kinetics & simulation for N, P, K constituent distribution of biofertilizer on plants like brinjal, chili has been studied and satisfactory validation has been developed.

5.0 REFERENCES:

1. Abera T., Tufa T., Midega T., Kumbi H., Tola, B. (2018). Effect of integrated inorganic and organic fertilizers on yield and yield components of Barley in Liben Jawi District. Int. J. Agron. 2018, 1–7.
2. Acevedo S.A., Carrillo A.J.D., Lopez E.F., Tovar C.D.G. (2021). Recovery of banana waste loss from production and processing: A contribution to a circular economy. J.Mol. 26(17).
3. Adams R., Besant J., Phelps R. (2006) Innovation Management Measurement: A review. Bus Corpo Fin Val.
4. Alvarez C.E., Garcia C., Carracedo A. (1988). Soil fertility and mineral nutrition of an organic banana plantation in Tenerife. Biol. Agric. Hortic. 5, 313–323
5. Chakraborty P., Banerjee S., Kumar S., Sadhukhan S., Halder G. (2018). Elucidation of ibuprofen uptake capability of raw and steam activated biochar of Aegle marmelos shell: isotherm, kinetics, thermodynamics and cost estimation. Process Saf. Environ. Protect. 118, 10–
6. Dhal B., Thatoi H., Das N., Pandey B. (2013). Chemical and microbial remediation of hexavalent chromium from contaminated soil and mining/metallurgical solid waste: a review. J. Hazard. Mater. 250, 272–291.
7. Gao D., Wang, P., Shi, J., Li, F., Li, W., Lyu, B., Ma, J. (2019). A green chemistry approach to leather tanning process: cage-like octa (aminosilsesquioxane) combined with Tetrakis (hydroxymethyl) phosphonium sulfate. J. Clean. Prod. 229, 1102–1111.
8. Karungi J., Ekbom B., Kyamanywa S. (2006). Effects of organic versus conventional fertilizers on insect pests, natural enemies and yield of Phaseolus vulgaris. Agric., Ecosyst. Environ. 115, 51–55.

9. Kolomaznik K., Adamek M., Andel I., Uhlirova M. (2008). Leather waste—potential threat to human health, and a new technology of its treatment. J. Hazard. Mater. 160, 514–520.
10. Lima de Oliveira D.Q., Gomes Carvalho K.T., Ribeiro Bastos A.R., Alves de Oliveira L. C., de Sa e Melo Marques J.J.G., de Melo Pereira do Nascimento R.S. (2008). Use of leather industry residues as nitrogen sources for elephant grass. Rev. Bras. Cienc. Solo 32, 417–424.
11. Lin W., Lin M., Zhou H., Wu H., Li Z., Lin W., 2019. The effects of chemical and organic fertilizer usage on rhizosphere soil in tea orchards. PloS One 14, e0217018.
12. Li X., Han X., Li H., Song C., Yan J., Liang Y. (2012). Soil chemical and biological properties affected by 21-year application of composted manure with chemical fertilizers in a Chinese Mollisol. Can. J. Soil Sci. 92, 419–428.
13. Majee S., Halder G., Mandal D.D., Tiwari O.N., Mandal T. (2021) Transforming wet blue leather and potato peel into an eco-friendly bioorganic NPK fertilizer for intensifying crop productivity and retrieving value-added chromium salts. J Haz Mat. 411, 125046.
14. Majee S., Sarkar K.K., Sarkhel R., Halder G., Mandal D.D., Rathinam N.K., Mandal T. (2023). Bioorganic fertilizer production from industrial waste and insightful analysis on release kinetics. J Env Manag. 325, 116378.
15. Kochakinezhad H., Peyvast G.-A., Kashi A.-K., Olfati J.-A., Asadi A. (2012). A comparison of organic and chemical fertilizers for tomato production. J. Organic Syst. 7, 14–25.
16. Kolomaznik K., Adamek M., Andel I., Uhlirova M. (2008). Leather waste—potential threat to human health, and a new technology of its treatment. J. Hazard. Mater. 160, 514–520.
17. Langrrd S. (1990). One hundred years of chromium and cancer: a review of epidemiological evidence and selected case reports. Am. J. Ind. Med. 17, 189–214.
18. Lloyd D.J., Gatehouse B.M. (1973) Dalton Transactions, Inorganic Chemistry. J. Che Soc. 70.
19. Majee S., Halder G., Mandal D.D., Tiwari O.N., Mandal T. (2021b). Transforming wet blue leather and potato peel into an eco-friendly bioorganic NPK fertilizer for intensifying crop productivity and retrieving value-added recyclable chromium salts. J. Hazard Mater. 411, 125046.
20. Malinowski P., Biskuspi A., Glowinski J. (2007). Preparation methods of calcium sulphate and urea adduct. Pol J. Che Tech. 9(4), 111-114.
21. Mertz W. (1969). Chromium occurrence and function in biological systems. Physio Rev. 49,163–239.
22. Mitter, E. K., Tosi, M., Obregon, D., Dunfield, K. E., and Germida, J. J. (2021). Rethinking crop nutrition in times of modern microbiology: innovative biofertilizer technologies. Front. Sustain. Food Syst. 5, 29.
23. Nogueira F.G.E., do Prado N.T., Oliveira L.C.A., Bastos A.R.R., Lopes J.H., de Carvalho J.G. (2010). Incorporation of mineral phosphorus and potassium on leather waste (collagen): a new collagen NPK-fertilizer with slow liberation. J. Hazard. Mater. 176, 374–380.
24. Parmar H.Y., Chakraborty H. (2016). Effect of siderophore on plant growth promotion. Int J. Appl. Pure Sci. Agric. 2, 60–68.
25. Patel D.H, Lakdawala M.M (2014). Study of soil's nature by pH and soluble salts through EC of Kalol-Godhra taluka territory. Der Chemica Sinica, 5(2), 1-7.
26. Park H.J., Byun M.K., Kim H.J., Ahn C.M., Rhee C.K., Kim K., Kim B.Y., Bae H.W., Yoo K.H. (2018). Risk for asthma exacerbation in Korean adults. J. All Asth Immun. 29(14).
27. Sharma A., Mittra B. (1991). Effect of different rates of application of organic and nitrogen fertilizers in a rice-based cropping system. J. Agric. Sci. 117, 313–318.
28. Shekhawat K., Chatterjee S., Joshi B. (2015). Chromium toxicity and its health hazards. Int. J. Adv. Res. 3, 167–172.
29. Singh A., Kumar V. (2022) Liquid Biofertilizers. Int Fert Tech. 17(02).

30. Suryanto A., Hamid A., Damaiyanti D.R.R. (2017) Effectiveness of Biofertilizer on growth and productivity of eggplant (*Solanum melangona L.*). J Adv. Agri Tech. 4.
31. Tasca, A.L., Puccini, M. (2019). Leather tanning: Life cycle assessment of retanning, fat liquoring and dyeing. J. Clean. Prod. 226, 720–729.23.
32. WHO (World Health Organization) (1996). Trace elements in human nutrition and health (A report of a re-evaluation of the role of trace elements in human health and nutrition). Geneva.
33. Zaini H.M., Roslan J., Sallah S., Munsu E., Sulaiman N.S., Pindi W. (2022). Banana peels as a bioactive ingredient and its potential in the food industry. J. Func Foods. 92, 105054.

105c Multicomponent Aqueous Phase Adsorption Equilibria of Organic Acids on Ion-Exchange Resin

Usman Hamid, Cheng-Ju Hsieh, Haoran Wu, Sadia Saberin, Nooram Anjum, Lauren Valentino, Meltem Urgun-Demirtas, Chau-Chyun Chen

Thermal separation of valuable organic acids from biomass fermentation broth requires high investment cost and high energy consumption. Adsorptive separation is advantageous due to its high efficiency, high selectivity, and reduced energy requirement as compared to thermal separation. To investigate adsorptive separation of organic acids from fermentation broth, systematic aqueous phase adsorption equilibrium measurements are carried out with the unary, binary, and ternary systems of acetic acid, butyric acid, and lactic acid on a commercial ion-exchange resin at different solution concentrations and pH values at room temperature. The incorporation of physical property and thermodynamic data into process simulations is crucial for evaluation of adsorbents and improvement of separations processes. Thermodynamic modeling of the adsorption equilibrium data is performed with a newly developed generalized Langmuir (gL) isotherm model for aqueous phase adsorption equilibrium. Taking into account competitive adsorption of adsorbates, thermodynamic, and solution chemistry of organic acids in aqueous solution of different pH, the model precisely captures the unary, binary, and ternary aqueous phase adsorption equilibria experimental data. The gL isotherm representations are also compared against those with conventional Overloading model and Ideal Dilute Solution Theory. The gL isotherm model provides a thermodynamically consistent foundation for multicomponent aqueous phase adsorption equilibria of organic acids in support of subsequent process simulation, development, and design of adsorptive separation units for organic acid separation from fermentation broth.

AUTHOR INDEX